16th Meeting on the Mathematics of Language (MOL 2019)

Toronto, Canada
18-19 July 2019

ISBN: 978-1-5108-9094-7

Printed from e-media with permission by:

Curran Associates, Inc.
57 Morehouse Lane
Red Hook, NY 12571

Some format issues inherent in the e-media version may also appear in this print version.

Copyright© (2019) by the Association for Computational Linguistics
All rights reserved.

Printed by Curran Associates, Inc. (2019)

For permission requests, please contact the Association for Computational Linguistics
at the address below.

Association for Computational Linguistics
209 N. Eighth Street
Stroudsburg, Pennsylvania 18360

Phone: 1-570-476-8006
Fax: 1-570-476-0860

acl@aclweb.org

Additional copies of this publication are available from:

Curran Associates, Inc.
57 Morehouse Lane
Red Hook, NY 12571 USA
Phone: 845-758-0400
Fax: 845-758-2633
Email: curran@proceedings.com
Web: www.proceedings.com

MOL 2019

The 16th Meeting on the Mathematics of Language

Proceedings of the Conference

18–19 July, 2019
University of Toronto
Toronto, Canada

UNIVERSITY OF
TORONTO

©2019 The Association for Computational Linguistics

Association for Computational Linguistics (ACL)
209 N. Eighth Street
Stroudsburg, PA 18360
USA
Tel: +1-570-476-8006
Fax: +1-570-476-0860
acl@aclweb.org

Introduction

These are the proceedings of the 16th Meeting on the Mathematics of Language (MOL 2019), held at the University of Toronto, on July 18–19, 2019.

The volume contains ten regular papers, which have been selected from a total of eighteen submissions, using the EasyChair conference management system.

The conference benefited from the financial support of the University of Toronto and of the Natural Sciences and Engineering Research Council of Canada, which we gratefully acknowledge.

Last but not least, we would like to express our sincere gratitude to all the reviewers for MOL 2019 and to all the people who helped with the local organization.

Philippe de Groote, Frank Drewes, and Gerald Penn (editors)

Program Chairs:

Philippe de Groote, INRIA Nancy - Grand Est (France)
Frank Drewes, Umeå University (Sweden)

Organizing Chair:

Gerald Penn, University of Toronto (Canada)

Program Committee:

Henrik Björklund, Umeå University (Sweden)
David Chiang, University of Notre Dame (USA)
Alexander Clark, King's College London (UK)
Valeria de Paiva, Nuance Communications (USA)
Carlos Gómez-Rodríguez, University of A Coruña (Spain)
Jeffrey Heinz, Stony Brook University (USA)
Makoto Kanazawa, Hosei University (Japan)
Greg Kobele, Leipzig University (Germany)
Marco Kuhlmann, Linköping University (Sweden)
Giorgio Magri, CNRS (France)
Andreas Maletti, Leipzig University (Germany)
Jens Michaelis, Bielefeld University (Germany)
Glyn Morrill, Universitat Politècnica de Catalunya (Spain)
Larry Moss, Indiana University, Bloomington (USA)
Reinhard Muskens, Tilburg University (The Netherlands)
Jim Rogers, Earlham College (USA)
Mehrnoosh Sadrzadeh, Queen Mary University of London (UK)
Anssi Yli-Jyrä, University of Helsinki (Finland)

Invited Speakers:

Matilde Marcolli, California Institute of Technology (USA)
Rohit Parikh, City University of New York (USA)

Table of Contents

Parsing Weighted Order-Preserving Hyperedge Replacement Grammars

Henrik Björklund
Dept. of Computing Science
Umeå University (Sweden)
henrikb@cs.umu.se

Frank Drewes
Dept. of Computing Science
Umeå University (Sweden)
drewes@cs.umu.se

Petter Ericson
Dept. of Computing Science
Umeå University (Sweden)
pettter@cs.umu.se

Abstract

We introduce a weighted extension of the recently proposed notion of order-preserving hyperedge-replacement grammars and prove that the weight of a graph according to such a weighted graph grammar can be computed uniformly in quadratic time (under assumptions made precise in the paper).

1 Introduction

The hyperedge-replacement grammar (HRG) is a well-studied formalism for describing graph languages; see, e.g., (Bauderon and Courcelle, 1987; Habel and Kreowski, 1987; Habel, 1992; Drewes et al., 1997). As argued by Jones et al. (2012), Koller (2015), and Groschwitz et al. (2015) it is also a promising candidate for modelling semantic representations of natural language such as Abstract Meaning Representation (AMR, see Banarescu et al. (2013)). However, HRGs overshoot the mark in that parsing with respect to them is computationally too expensive. Further, HRGs can express intricate structural properties whose complexity is far beyond what seems to be required to describe practically relevant languages of semantic graphs such as AMR. For example, as argued by Chiang et al. (2018) it suffices if the path languages of such graph languages are regular languages. In contrast, HRGs easily give rise to even non-context-free path languages. Thus, from both perspectives less powerful special cases should be sought if this helps to cut down on parsing complexity. Recently, such a restriction, called order preservation, was proposed and studied in (Björklund et al., 2016; Björklund et al., 2017; Björklund et al., 2018).

The present article builds upon the *order-preserving HRGs* (OPHGs) of Björklund et al. (2018), where it was shown that parsing for OPHGs is efficient, requiring polynomial time even in the uniform case i.e. when the grammar is considered to be part of the input. Here, we define a weighted version of OPHGs, and extend the results of Björklund et al. (2018) to show that when the weights are taken from a commutative semiring, we can efficiently compute the weight assigned by an OPHG to any input graph. This is an important feature since applications such as semantic modelling require ways to quantify the well-formedness of a generated graph.

While providing a notion of grammars with weights may appear to be a simple task as one only has to assign weights to the rules, doing so in a meaningful way for unrestricted HRGs is actually not simple at all. The reason is that the weights of different derivation trees generating the same graph should be summed up to obtain the weight of the graph. However, if a right-hand side of a rule has nontrivial automorphisms that interchange two or more nonterminal hyperedges, one gets spuriously distinct derivation trees that should intuitively be considered identical. At the very least, this complicates uniform parsing as it requires to preprocess the rules to detect the automorphisms of their right-hand sides, a task for which no polynomial solution is known.

In OPHGs, only the right-hand sides of so-called duplication rules have nontrivial automorphisms, and those do not require preprocessing. These rules correspond to associative and commutative operations, which we propose to take special care of in the computation of weights by using a type of reduced derivation trees introduced for the same purpose by Courcelle (1991a); see also Courcelle and Engelfriet (2012). In these derivation trees, some nodes have a set of children, while others have them ordered in a list. After this, we show how weights can efficiently be computed, and prove the correctness of the algorithm.

Related work. Another type of restricted HRGs for semantic modelling was proposed by Chiang et al. (2013), together with a parsing algorithm and a detailed complexity analysis. The complexity is, however, exponential even in the non-uniform case. In particular, it is exponential in the maximum degree of nodes in the input graph. The same holds for the parsing algorithm for *regular graph grammars* presented by Gilroy et al. (2017). We also mention that another technique for efficient HRG parsing was resently developed by Drewes et al. (2015, 2017).

2 Preliminaries

The set of non-negative integers is $\mathbb{N}$, and $[k] = \{1, \ldots, k\}$. For a set S, S^* is the set of strings over S, while $S^\circledast$ is the set of strings in S^* in which no element of S occurs twice. The empty string is ϵ, and we have $S^+ = S^* \setminus \epsilon$ and $S^\oplus = S^\circledast \setminus \epsilon$. The length of a string w is denoted $|w|$. We use the terms 'string' and 'sequence' interchangably. For a sequence $w = a_1 \cdots a_n$, every sequence $a_{i_1} \cdots a_{i_k}$ with $1 \leq i_1 < \cdots < i_k \leq n$ is a *subsequence* of w, and $[w]$ is the set $\{a_1, \ldots, a_n\}$.

2.1 Hypergraphs

We fix a disjoint, countably infinite supply LAB of labels, such that each $\sigma \in$ LAB has a rank $\mathrm{rank}(\sigma) \in \mathbb{N}$. A *hypergraph* is a structure $g = (V, E, \mathrm{lab}, \mathrm{att}, \mathrm{ext})$ where V and E are the (finite) sets of *nodes* and *hyperedges*, $\mathrm{lab} : E \to$ LAB is the *edge labelling*, $\mathrm{att} : E \to V^\oplus$ is the *edge attachment* with $|\mathrm{att}(e)| = \mathrm{rank}(\mathrm{lab}(e)) + 1$ for all $e \in E$, and $\mathrm{ext} \in V^\oplus$ is the sequence of *external nodes*.

From now on, we simply call hypergraphs graphs, and hyperedges edges. We use the graph as a subscript to identify its components. E.g., E_g refers to the set of edges of g. For an edge $e \in E_g$ with $\mathrm{att}_g(e) = v_0 \cdots v_k$, we say that $\mathrm{src}_g(e) = v_0, \mathrm{tar}_g(e) = v_1 \cdots v_k$, and name these the *source* and sequence of *targets*, respectively. Similarly, for $\mathrm{ext}_g = v_0 \cdots v_l$, we say that $v_0 = \dot{g}$ is the *source* of the graph, and $v_1 \cdots v_l = g_{\bullet\bullet}$ its sequence of *targets*. In this paper, we require all targets of a graph to be leaves, i.e. $\mathrm{src}_g(e) \notin [g_{\bullet\bullet}]$ for all $e \in E_g$. For a graph g, $\mathrm{rank}(g) = |g_{\bullet\bullet}|$, and for an edge e, $\mathrm{rank}(e) = \mathrm{rank}(\mathrm{lab}_g(e)) = |\mathrm{tar}_g(e)|$. Graphs g, h are *isomorphic*, denoted $g \equiv h$, if they are equal up to a bijective renaming of nodes and edges.

For $a \in$ LAB with $\mathrm{rank}(a) = k$, $a^\bullet$ denotes the graph $(\{v_0, \ldots, v_k\}, \{e\}, (e \to a), (e \to v_0 \cdots v_k), (v_0 \cdots v_k))$, i.e. the graph of one a-labelled edge of the proper rank, with all its attached nodes external.

An alternating sequence $v_1 e_1 \ldots v_k e_k$ of nodes and edges is a *path* in g from v_1 to e_k if $\mathrm{src}_g(e_i) = v_i$ and $v_{i+1} \in [\mathrm{tar}_g(e_i)]$, for each $i \in [k]$. We may optionally terminate the path at v_{k+1} instead of e_k. In either case, the path *passes* all nodes and edges v_i and e_i for $i \in [k]$. If $v_1 = \dot{g}$, it is a *source path*. A node v or edge e is *reachable from s* (in g) if there is a path in g from s to v (e). A node or edge is *reachable in g* if there is a source path to it.

2.2 Hyperedge replacement

Consider graphs h, f, and an edge $e \in E_h$ such that $\mathrm{rank}(e) = \mathrm{rank}(f)$, $V_h \cap V_f = [\mathrm{att}_h(e)]$, and $\mathrm{att}_h(e) = \mathrm{ext}_f$. Then we can use *hyperedge replacement* to obtain the graph $g = h[\![e : f]\!]$, *substituting f for e in h*, where $g = ((V_h \cup V_f), (E_h \cup E_f) \setminus \{e\}, \mathrm{att}_g, \mathrm{lab}_g, \mathrm{ext}_h)$ with

$$\mathrm{att}_g(e') = \begin{cases} \mathrm{att}_f(e') & \text{if } e' \in E_f \\ \mathrm{att}_h(e') & \text{if } e' \in E_h \setminus \{e\} \end{cases}$$

and

$$\mathrm{lab}_g(e') = \begin{cases} \mathrm{lab}_f(e') & \text{if } e' \in E_f \\ \mathrm{lab}_h(e') & \text{if } e' \in E_h \setminus \{e\}. \end{cases}$$

Clearly, if $\mathrm{rank}(e) = \mathrm{rank}(f)$ then we can always choose isomorphic copies of h and f, renaming nodes in such a way that $h[\![e : f]\!]$ is defined. We will generally not make note of this, to avoid irrelevant technicalities.

For the case where $g = h[\![e : f]\!]$ and $i = g[\![e' : j]\!]$ with $e' \notin E_f$, we write $i = h[\![e : f, e' : j]\!]$, and similarly for a larger number of replacements.

We divide LAB into two subsets TLAB and NLAB of *terminals* and *nonterminals*, and accordingly call edges terminal and nonterminal ones. We sometimes shorten the expressions further to just "terminals" and "nonterminals".

2.3 Hyperedge replacement grammars

A *hyperedge replacement grammar* (HRG) $G = (\Sigma, N, S, R)$ consists of a *terminal alphabet* $\Sigma \subset$ TLAB, a *nonterminal alphabet* $N \subset$ NLAB, an *initial nonterminal* $S \in N$, and a set R of *(HR) rules* form $A \to f$, where $A \in N$ and f is a graph over $\Sigma \cup N$ with $\mathrm{rank}(A) = \mathrm{rank}(f)$. If f has ℓ nonterminal edges, we name them $\{e_1, \ldots, e_\ell\}$ and write $arity(A \to f)$ for ℓ.

Derivations in HRGs are context-free: Given a graph h, an edge $e \in E_h$ with $\mathrm{lab}_h(e) = A \in N$, and a rule $(A \to f) \in R$, we can *derive* the graph $g = h[\![e : f]\!]$ from h. We call this a *derivation step*, and denote it $h \to_{A \to f} g$. We also write more generally $h \to_G g$ for a derivation step using any rule in R. The reflexive and transitive closure of $\to_G$ is $\to_G^*$. The *language of G* is the set $\mathcal{L}(G)$ of all graphs g over TLAB such that $S^\bullet \to_G^* g$.

3 Order-Preserving Hyperedge Replacement Grammars

We now turn to order-preserving HRGs. The first ingredient is a condition called reentrancy preservation. Reentrancies are deeply entwined with the way we identify places in a graph that match the right-hand side of a given rule.

3.1 Reentrancies

Suppose we consider a subgraph h of a graph g as a candidate of a subgraph that may have been derived from a nonterminal e. If so, then $g = g'[\![e : h]\!]$ where, intuitively, g' is obtained from g by replacing h by e. To perform this backwards replacement, we have to determine which nodes of h are its external nodes, i.e., which ones are to be attached to e. By the very definition of hyperedge replacement, a node of h that is external in g or has an attached edge not belonging to h, must be in $[\mathrm{att}_{g'}(e)]$ (but not generally vice versa). In particular, all nodes in h that can be reached from $\dot{g}$ without passing a node in h must be in $[\mathrm{att}_{g'}(e)]$. The notion of reentrant nodes to be defined now serves to turn this inclusion into an equality (once we add $[\mathrm{ext}_g] \cap V_h$ to this set) in the case where h is rooted at some node or edge x of g.

Intuitively, the *reentrant nodes* of a node or edge x in a graph g are the first descendants of x that can also be reached on a path that avoids x. As the external nodes of a right-hand side of an HR rule are the ones that, after the replacement, are reachable from "outside" the subgraph, we also consider them as reentrant. The graph delineated by x and its reentrant nodes is the *subgraph rooted at x*.

Let us have a look at a simple example before defining the notion of reentrant nodes formally. The graph in Figure 1 is single-rooted, with r the root node. The reentrant nodes of r is the set of external targets (i.e. x_1, x_2 and x_3), and these are also the reentrant nodes of the edge e sourced at r.

For the edge marked f, x_2 is a reentrant node, and so is v_1 and v_2, as v_2 is reachable through the path rei_1gv_2 that avoids f, and v_1 likewise is reachable by the path $rei_1gi_2hv_1$, also avoiding f. For f', the set of reentrant nodes is $\{v_1, v_3\}$, as v_3 is also a direct target of f, making it reachable on the path rei_3fv_3 that avoids f'.

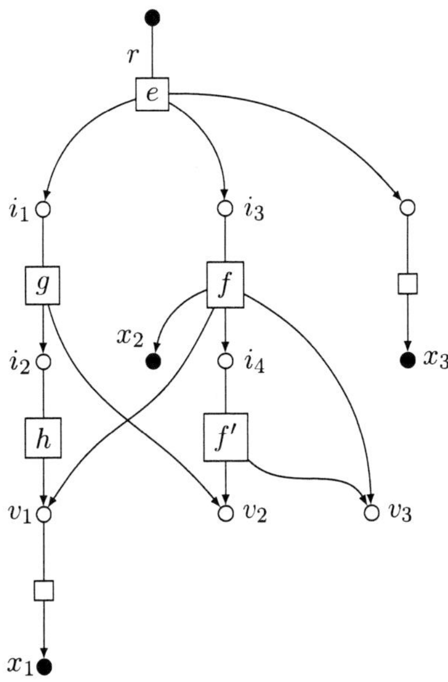

Figure 1: An example graph for reentrancies.

Definition 3.1 (Reentrant node). *Given a graph g and $E \subset E_g$, let $\mathrm{TAR}_g(E)$ be the union of all sets of targets of edges in E, i.e. $\bigcup_{e \in E}[\mathrm{tar}_g(e)]$.*

Further, for $x \in V_g \cup E_g$, let $\hat{x}$ be x if $x \in V_g$, and $\mathrm{src}_g(x)$ if $x \in E_g$. Now, let E_g^x be the set of all edges $e \in E_g$ such that all source paths to e pass x.[1] Then the set of reentrant nodes of x in g is

$$\mathrm{reent}_g(x) = \begin{aligned}&(\mathrm{TAR}_g(E_g^x) \setminus \{\hat{x}\}) \cap \\ &(\mathrm{TAR}_g(E_g \setminus E_g^x) \cup [\mathrm{ext}_g]).\end{aligned}$$

Definition 3.2 (Rooted subgraph). *Given a graph g with $x \in V_g \cup E_g$, the* subgraph $g{\downarrow}_x$ rooted at x *is a graph h such that $E_h = E_g^x$, $V_h = \{\hat{x}\} \cup \mathrm{TAR}_g(E_h)$, att_h and lab_h are the appropriate restrictions of att_g and lab_g, respectively, and ext_h is $\hat{x}$ followed by $\mathrm{reent}_h(x)$ in some order.*

Rooted subgraphs are strictly nested, which is proved by Björklund et al. (2018) in the form of the following lemma (where $\sim$ is isomorphy modulo the order of $g_{..}$):

---

[1] Note that if x is not reachable in g, $E_g^x = \emptyset$

Lemma 3.3 (Lemma 3.4 in (Björklund et al., 2018)). *Let g be a graph, $h = g{\downarrow}_x$ for some $x \in V_g \cup E_g$. Then $h{\downarrow}_y \sim g{\downarrow}_y$ for all $y \in (V_h \cup E_h) \setminus [\mathrm{ext}_h]$*

3.2 Reentrancy Preservation

Reentrancy preservation formalizes the property that, given a graph h and some edge $e \in E_h$ with $\mathrm{lab}_h(e)$, we can replace e by some graph f according to a rule $A \to f$ without affecting the sets $\mathrm{reent}_g(x)$ for $x \in V_h \cup V_f$.

We achieve this by restricting our grammars to two types of rules, namely *duplication rules* and *deep rules*. Rules of these two kinds are called *reentrancy preserving*. To define duplication rules, consider a graph

$$f = (\{v_0, \ldots, v_n\}, \{e_1, e_2\}, \mathrm{att}, \mathrm{lab}, \mathrm{ext}),$$

where $\mathrm{att}(e_1) = v_0 \cdots v_n = \mathrm{att}(e_2)$, $\mathrm{lab}(e_1) = \mathrm{lab}(e_2) \in \mathrm{NLAB}$, and ext is a subsequence of $\mathrm{att}(e_1)$ starting with v_0. If $|\mathrm{ext}| < n$ then f (and every graph isomorphic to f) is a *twin*, and if $|\mathrm{ext}| = n$ then it is a *clone*. A rule $A \to f$ is a *twin rule* if f is a twin and a *clone rule* if f is a clone with $\mathrm{lab}(e_1) = \mathrm{lab}(e_2) = A$. A *duplication rule* is either a clone or a twin rule.

A rule $A \to f$ is a *deep rule* if f fulfills the following conditions:

- $V_f \neq [\mathrm{ext}_f]$,

- all nodes in V_f are reachable from $\dot{f}$ and have out-degree ≤ 1, and

- for every nonterminal edge e, $\mathrm{reent}_f(e) = [\mathrm{tar}_f(e)]$.

A HRG is reentrancy preserving if it has only reentrancy-preserving rules. We note here that Björklund et al. (2018) also permits chain rules, i.e. rules that only change the label of an edge from one nonterminal to another nonterminal, and thus violate the first condition above. In the present paper we exclude them because they can result in an infinite number of derivations of a given graph, thus making it in general unreasonable to associate a weight with such a graph.[2]

Later on, we will also need the following generalization of duplication rules to the case where $\ell + 1$ copies of a nonterminal edge are created: given any duplication rule $r = (A \to f)$ and some $\ell \geq 1$, we denote by r^ℓ the rule $A \to f'$, where f' is obtained from f by replacing its two nonterminals by $\ell + 1$ copies. Thus, $r^1 = r$.

Lemma 3.4 (Björklund et al. (2018) adapted). *Let $g \in \mathcal{L}(G)$ for some reentrancy-preserving HRG G. There is a quadratic algorithm that computes, for every $x \in V_g \cup E_g$, the set $\mathrm{reent}_g(x)$, and thus the subgraph $g{\downarrow}_x$.*

3.3 Ordering nodes

Reentrancy preservation allows us to pinpoint the subgraphs that may have been generated by a specific nonterminal, but as shown by Björklund et al. (2016), this is not sufficient to achieve efficient parsing, as needing to guess the order of targets in subgraphs $g{\downarrow}_x$ may still cause NP-hardness. Thus, we require a way to determine the order of nodes, in particular reentrant nodes. This requires an ordering relation that can be efficiently computed, and fulfils some basic requirements, and a set of reentrancy-preserving rules that additionally *preserves that order*. Formally:

Definition 3.5 (Suitable order). *For a set $\mathcal{G}$ of graphs, a* suitable family of orders *is a family $(\preceq_g)_{g \in \mathcal{G}}$ of binary relations $\preceq_g \subseteq V_g \times V_g$ such that*

- *for all $A \in \mathrm{LAB}_N$, $A^{\bullet}_{\cdot\cdot}$ is ordered by $\preceq_{A^\bullet}$ and*

- *if $i : g \to h$ is an isomorphism and $u, v \in V_g$, then $u \preceq_g v$ iff $i_V(u) \preceq_h i_V(v)$.*

Definition 3.6 (Order preservation). *A reentrancy-preserving set R of HR rules* preserves *a suitable family of orders $\preceq = (\preceq_g)_{g \in \mathcal{G}}$ if, for all $g = h[e : f]$ with $g, h, f \in \mathcal{G}$, $e \in E_h$, and $\mathrm{lab}_h(e) \to f \in R$, we have $\preceq_g|_{V_h} = \preceq_h$ and $\preceq_f|_{V_f} = \preceq_f$.*

An order-preserving HRG (OPHG) *is a reentrancy preserving HRG (Σ, N, S, R) together with a suitable family $\preceq$ of orders preserved by R.*

4 Weighted Order-Preserving HR Grammars

We now add weights – taken from some semiring – to order-preserving HR grammars. For this, and throughout the rest of this paper, let $\mathcal{S} = (S, +, \cdot, 0, 1)$ be a commutative semiring, meaning that $(S, +, 0)$ and $(S, \cdot, 1)$ are two monoids over the domain S such that $\cdot$ distributes over $+$. Thus, spelled out in detail, $+$ and $\cdot$ are binary operations on S such that

---

[2]To allow for chain rules, one may require the semiring to be complete, i.e., to have infinite sums. We do not pursue this possibility here.

- 1 is the identity element for $\cdot$

- 0 is the identity element for $+$ and the absorbing one for $\cdot$,

- $+$ and $\cdot$ are commutative, and

- $\cdot$ distributes over $+$.

As usual, for every $a \in S$ we let $a^0 = 1$ and $a^{n+1} = a \cdot a^n$ for all $n \in \mathbb{N}$.

Examples of well-known semirings are the Boolean semiring, the real numbers with addition and multiplication, the *tropical semiring* consisting of the positive real numbers extended by ∞ with minimum and addition, and the Viterbi semiring over $[0, 1]$ in which multiplication is as usual and addition is maximum. The latter is used in natural language processing to compute the likelihood of the most probable derivation. See (Goodman, 1999) for more information on the use of semirings in natural language parsing.

A weighted OPHG computes a graph series, i.e. a mapping of graphs to S. As usual, this is achieved by assigning weights to rules.

Definition 4.1 (weighted OPHG). *A weighted OPHG $G = (\Sigma, N, S, R, \omega)$ (over S) consists of an OPHG (Σ, N, S, R) and a weight assignment $\omega \colon R \to S$.*

Informally speaking, if several distinct derivations can produce the same graph, we sum up the weights of the individual derivations to obtain the weight of the graph. The weight for a single derivation is the product of the weights of all the rules applied.

It is inconvenient to formalise this based on the derivations themselves because, just as in the case of ordinary context-free grammars, derivations may differ only in the order in which nonterminals are replaced, which yields distinct derivations that should be considered equivalent. A standard technique to solve this problem is to consider derivation trees instead of derivations. We can mostly use this standard technique, but we propose to take into account the fact, mentioned in the introduction, that each duplication rules has a nontrivial automorphism that interchanges the nonterminals in its right-hand side. Hence, these nonterminals are indistinguishable. Moreover, if the rule is a clone rule, then applying it to any of the nonterminals in its right-hand side yields three indistinguishable nonterminals in two different ways.

In general, suppose that a nonterminal is cloned ℓ times, yielding $\ell + 1$ copies which are then further derived into graphs $g_0, \ldots, g_\ell$ of weights $w_0, \ldots, w_\ell$. Then the clones can be derived by C_ℓ different derivation trees, where C_ℓ is the ℓ-th Catalan number (i.e., the number of binary trees with $\ell + 1$ leaves). The resulting nonterminals $e_0, \ldots, e_\ell$ can be derived into the graphs $g_0, \ldots, g_\ell$ in any order, all leading to the same result. This yields $\ell! C_\ell$ distinct derivations, all generating the same graph g which consists of $g_0, \ldots, g_\ell$ fused at their external nodes. The weight of g would thus be $w^\ell \sum_{j=1}^{\ell! C_\ell} \prod_{i=0}^\ell w_i$, where w is the weight of the cloning rule. While there is nothing wrong with this in principle, the fact that we only allow for this particular type of cloning rule implies that there would be no way to avoid the sum by writing the rules of the grammar in a different way. Further, since the number of terms summed up depends on ℓ, it cannot in general be compensated for by reducing the weights of rules. We expect this to be a limiting factor in applications, and thus propose to represent a ℓ-fold cloning as an unordered node of rank $\ell + 1$ in the derivation tree, leading to the weight $w^\ell \prod_{i=0}^\ell w_i$.

Let us begin the process of making these notions more precise by recalling the notions of *shallow graphs* and *siblinghoods* from (Björklund et al., 2018).

Definition 4.2. *A graph g is shallow if $\dot{g} = \mathrm{src}_g(e)$ for all $e \in E_g$. A siblinghood in g is a set $\mathrm{Sib} \subseteq E_g$ such that $|\mathrm{Sib}| \geq 2$ and $\mathrm{tar}_g(e) = \mathrm{tar}_g(e')$ for all $e, e' \in \mathrm{Sib}$. We denote $\mathrm{tar}_g(e)$, $e \in \mathrm{Sib}$, by $\mathrm{tar}_g(\mathrm{Sib})$, and let $g(\mathrm{Sib}) = (\{\dot{g}\} \cup [\mathrm{tar}_g(\mathrm{Sib})], \mathrm{Sib}, \mathrm{att}_g|_{\mathrm{Sib}}, \mathrm{lab}_g|_{\mathrm{Sib}}, \mathrm{tar})$, where tar is the subsequence of $\mathrm{tar}_g(\mathrm{Sib})$ of nodes that are external in g or targets of edges outside of Sib, i.e. that belong to the set*

$$\mathrm{TAR}_g(\mathrm{Sib}) \cap (\mathrm{TAR}_g(E_g \setminus \mathrm{Sib}) \cup [g_.]).$$

For siblinghoods $\mathrm{Sib}, \mathrm{Sib}'$, we let $\mathrm{Sib} \leq \mathrm{Sib}'$ if $\mathrm{tar}_g(\mathrm{Sib})$ is a subsequence of $\mathrm{tar}_g(\mathrm{Sib}')$. A siblinghood of g is prime if it is maximal with respect to both $\leq$ and set inclusion.

From now on, we shall for technical simplicity assume that the considered OPHG G contains exactly one clone rule for every $A \in N$. This is not a restriction because the definition of the weight of derived graphs to be given below ensures that any number of clone rules for the same nonterminal can be replaced by a single clone rule whose weight is the sum of the weights of the individual rules. In particular, if there is no clone rule for A, this has the same effect as a single clone rule of weight 0. The weight of the unique clone rule for $A \in N$ is denoted by $\omega(A)$, and we write $\to_{cl}$ for the derivation relation that exclusively uses clone rules, i.e. $g \to_{cl}^* g'$ if g' is obtained from g by cloning nonterminal edges.

The following is essentially Lemma 5.3 of (Björklund et al., 2018):

Lemma 4.3. *Let $A \in N$ and let g be a shallow graph over N with $|E_g| \geq 2$.*

- *If $A^\bullet \to^+ g$, then for every prime sibling-hood Sib of g we either have $g = g(\mathrm{Sib})$ and $A^\bullet \to_{cl}^+ g$, or $A^\bullet \to^* h \to h[\![e : f]\!] \to_{cl}^* h[\![e : f']\!] = g$ where $\mathrm{lab}_h(e) \to f$ is a twin rule and $g(\mathrm{Sib}) = f'$.*

- *Up to reordering of derivation steps, the derivations of these forms are the only ones deriving g from $A^\bullet$.*

Hence, a derivation of a shallow graph can be broken down into an initial series of clonings followed by iterated sub-derivations each consisting of an application of a twin rule $A \to f$ and any number of clonings of the two nonterminal edges e_1, e_2 of f. Note that the result of each such sub-derivation depends only on $A \to f$ and the number of clonings since $\mathrm{att}_f(e_1) = \mathrm{att}_f(e_2)$. Therefore, the following definition of derivation trees uses trees in which the nodes that correspond to derivations of siblinghoods are unordered and unranked. For a tree consisting of a root labelled a and subtrees $t_1, \ldots, t_\ell$, we write $a[t_1, \ldots, t_\ell]$ or $a\langle t_1, \ldots, t_\ell\rangle$ depending on whether $t_1, \ldots, t_\ell$ is to be interpreted as an ordered or unordered list (or a multiset), respectively. We write $a(t_1, \ldots, t_\ell)$ to denote a tree in which the first level of children can be either ordered or unordered.

Definition 4.4 (derivation tree). *For a weighted OPHG $G = (\Sigma, N, S, R, \omega)$ and $A \in N$, the set of all A-derivation trees is the smallest set of trees t belonging to one of the following three types:*

(1) $t = r[t_1, \ldots, t_\ell]$ for a deep rule $r = (A \to f) \in R$ such that $arity(A \to f) = \ell$, and t_i is a $\mathrm{lab}_f(e_i)$-derivation tree for every $i \in [k]$.

(2) $t = r^\ell \langle t_1, \ldots, t_{\ell+1}\rangle$ for a clone rule $A \to f$, where $\ell \geq 1$ and, for every $i \in [\ell + 1]$, the subtree t_i is an A-derivation tree that is not of type (2).

(3) $t = r^\ell \langle t_1, \ldots, t_{\ell+1}\rangle$ for a twin rule $A \to f$, where $\ell \geq 1$ and, for every $i \in [\ell + 1]$, the subtree t_i is a $\mathrm{lab}_f(e_1)$-derivation tree that is not of type (2).

A more rigorous and complete treatment of various issues surrounding derivation trees of graph algebras with associative and commutative operations can be found in (Courcelle, 1991b).

We can *evaluate* a derivation tree to yield a graph g in the following way: Given a derivation tree $t = r(t_1, \ldots, t_\ell)$, $eval(t)$ is defined as the right-hand side f of r, with each successive nonterminal e_i replaced with the evaluation of the corresponding subtree of the derivation tree, i.e. $eval((A \to f)(t_1, \ldots, t_\ell)) = f[\![e_1 : eval(t_1), \ldots, e_\ell : eval(t_\ell)]\!]$. Given a graph g, we let $\mathrm{DT}_G(g)$ denote the set of all S-derivation trees such that $eval(t) \equiv g$.

We make the following observation, whose correctness follows from the context-freeness of hyperedge replacement.

Observation 4.5. *Let $G = (\Sigma, N, S, R, \omega)$, be an OPHG. Then it holds that*

$$\mathcal{L}(G) = \{ eval(t) \mid t \text{ is an } S\text{-derivation tree of } G\}.$$

Now, as mentioned, the weight of a graph is defined to be the sum of the weights of all its derivation trees:

Definition 4.6 (generated graph series). *Let $G = (\Sigma, N, S, R, \omega)$ be a weighted OPHG and $A \in N$.*

1. For every duplication rule $r = (A \to f) \in R$ and every $\ell \geq 1$, let $\omega(r^\ell) = \omega(r) \cdot \omega(\mathrm{lab}_f(e_1))^{\ell-1}$. (Note that r^ℓ corresponds to the application of r followed by $\ell - 1$ clonings of any of the two resulting nonterminal edges.)

2. The weight of an A-derivation tree $t = r(t_1, \ldots, t_\ell)$ ($\ell \in \mathbb{N}$) is defined inductively, as

$$\omega(t) = \omega(r) \cdot \prod_{i \in [k]} \omega(t_i).$$

3. The graph series $\omega_G \colon \mathcal{G}_\Sigma \to S$ generated by G is given by

$$\omega_G(g) = \sum_{t \in \mathrm{DT}_G(g)} \omega(t).$$

(The sum is finite, and thus well defined due to the commutativity of $+$.)

Note that given G, the language $\mathcal{L}(G)$ of G seen as an unweighted grammar, is a superset of the *support* of G, i.e. the set of all graphs g such that $\omega_G(g) \neq 0$.

5   Computing Weights

Our algorithm builds upon the unweighted parsing algorithm by Björklund et al. (2018). We store in each node and edge nothing more than an $|N|$-vector of weights, which is computed in very much the same way as the sets of nonterminals computed in (Björklund et al., 2018). We use the distributivity of multiplication over addition to keep our computations efficient (assuming efficient multiplication and addition).

The algorithm exploits Lemma 3.3, i.e. the property that the subgraphs $g{\downarrow}_x$ are strictly nested in all graphs derivable by an OPHL. Using this, it is possible to process the subgraphs of g in a tree-like "bottom-up" manner, marking each node and edge x with the set of all nonterminals that can generate $g{\downarrow}_x$, after all $g{\downarrow}_y$ properly contained in $g{\downarrow}_x$ have already been processed. Eventually, S belongs to the set which the node $\dot{g}$ is marked with if and only if $g \in \mathcal{L}(G)$.

Order preservation enters the picture as follows: every subgraph h of g which was derived from some nonterminal edge, is of the form $h = g{\downarrow}_x$ for some node or edge x of g. As shown by Björklund et al. (2018), order preservation guarantees that $h_{\bullet\bullet}$ is ordered by $\preceq_g$. Thus, in the algorithm only those subgraphs $g{\downarrow}_x$ are of interest for which the ordering of targets is uniquely determined by $\preceq_g$. From now on, we will thus assume that, whenever a subgraph $h = g{\downarrow}_x$ is constructed, the order of nodes in $h_{\bullet\bullet}$ is chosen according to $\preceq_g$.

To show how $\omega_G(g)$ can be computed, we describe two algorithms in one: the first computes the derivation trees of g whereas the second computes its weight by summing up over all the derivation trees. In the current paper, we mainly use the first algorithm as a tool to facilitate the correctness proof of the second. As a consequence, we do not present that first algorithm in a way which immediately yields an efficient algorithm, i.e., we only care for the efficiency of the second algorithm. The set of derivation trees computed by the first algorithm can, however, be represented in a compact fashion as a "packed forest", which is of independent usefulness and makes the algorithm efficient.

The main procedure of the algorithm computes, in the same bottom-up manner as in (Björklund et al., 2018), a set $\mathcal{D}_x(A)$ of A-derivation trees for each $x \in V_g \cup E_g$ and every $A \in N$. More precisely, $\mathcal{D}_x(A)$ is the set of all A-derivation trees of the input HRG G such that $A^\bullet \to_G^* g{\downarrow}_x$. As the correctness of this procedure was proved by Björklund et al. (2018) (though not explicitly in terms of derivation trees), all that remains to be shown is that the second version of the algorithm computes $\sum_{t \in \mathcal{D}_{\dot{g}}(S)} \omega(t)$ under the assumption that the first one is correct.

That second algorithm computes weights $\mathcal{W}_x(A)$ instead of the sets $\mathcal{D}_x(A)$, where $\mathcal{W}_x(A) = \sum_{t \in \mathcal{D}_x(A))} \omega(t)$. In the pseudocode, we always indicate the changes that must be made to obtain the second version by lines marked by "***alt:***". The line marked in this manner replaces its immediate predecessor. For sets of (derivation) trees $D_1, \ldots, D_\ell$ ($\ell \in \mathbb{N}$) and a rule r of arity ℓ, we furthermore write $r(D_1, \ldots, D_\ell)$ to denote the set

$$\{r(t_1, \ldots, t_\ell) \mid (t_1, \ldots, t_\ell) \in D_1 \times \cdots \times D_\ell\}$$

(i.e. we use that notation in both the ordered and unordered case).

A subroutine used by the algorithm is Algorithm 1, a modified version of the corresponding procedure in (Björklund et al., 2018). It takes as input a shallow graph h whose edges e are already assumed to be annotated with the respective sets $\mathcal{D}_e(A)$. The algorithm uses Lemma 4.3 in order to assemble – in a bottom-up manner over the prime siblinghoods of h – the set $\mathcal{D}_{\dot{h}}(A)$. In the algorithm we say that a duplication rule $A \to f$ of G *fits* a siblinghood $\mathrm{Sib} = \{s_1, \ldots, s_\ell\}$ of h if $f \equiv h(\{s_1, s_2\})$ when disregarding edge labels, and we denote f by $B^{\bullet\bullet}$ to indicate that the two edges in f carry the label B.

The reader should note that the result of Algorithm 1 does not depend on the choice of Sib because the prime siblinghoods $\mathrm{Sib}_1, \ldots, \mathrm{Sib}_k$ of h are pairwise disjoint and the replacement of $\mathrm{Sib} = \mathrm{Sib}_i$ by e does not affect the siblinghoods $\mathrm{Sib}_j, j \in [k] \setminus \{i\}$ (though it may of course create an additional prime siblinghood).

The main procedure of the parsing algorithm is shown in Algorithm 2. In its while loop, it repeatedly chooses an $x \in V_g \cup E_g$ for which the sets $\mathcal{D}_x(A)$ shall be computed, and calls $\textsc{Parse}_\mathrm{V}$ (Algorithm 3) or $\textsc{Parse}_\mathrm{E}$ (Algorithm 4) depending on whether $x \in V_g$ or $x \in E_g$.

The function $\textsc{Matching}$ used in line 4 of Al-

Algorithm 1 Computing Derivation Trees with Duplication Rules

---

1: **function** SHALLOWPARSE(set R of duplication rules, shallow annotated graph h with irrelevant edge labels)

2:     **while** $|E_g| > 1$ **do**

3:         **if** h does not contain a prime siblinghood **then**

4:             **return** $(A \mapsto \emptyset)_{A \in N}$

            alt: **return** $(A \mapsto 0)_{A \in N}$

5:         choose a prime siblinghood Sib $= \{s_1, \ldots, s_{\ell+1}\}$ ($\ell \geq 1$)

6:         replace Sib in h by a new edge e with $\mathrm{tar}_h(e) = h(\mathrm{Sib})$..

7:         **for** each $A \in N$ **do**

8:             $\mathcal{D}_e(A) \leftarrow \bigcup_{r \, = \, (A \to B^{\bullet\bullet}) \text{ fits Sib}} r^{\ell} \langle \mathcal{D}_{s_1}(B), \ldots, \mathcal{D}_{s_{\ell+1}}(B) \rangle$

            alt: $\mathcal{W}_e(A) \leftarrow \sum_{r \, = \, (A \to B^{\bullet\bullet}) \text{ fits Sib}} \omega(r^{\ell}) \cdot \prod_{i \in [\ell+1]} \mathcal{W}_{s_i}(B)$

9:     **return** $(A \mapsto \mathcal{D}_e(A))_{A \in N}$ where $\{e\} = E_h$

    alt: **return** $(A \mapsto \mathcal{W}_e(A))_{A \in N}$ where $\{e\} = E_h$

Algorithm 2 Computing Derivation Trees for Order-Preserving HR Grammars

---

1: **function** PARSE(order-preserving HR grammar $G = (\Sigma, N, S, R)$, graph $g \in \mathcal{G}_R$)

2:     $preProcess(g)$ $\triangleright$ Compute $\prec_g$ as well as all $g{\downarrow_x}$ for all $x \in V_g \cup E_g$

3:     **for** $x \in V_g \cup E_g$ **do**

4:         **if** $g{\downarrow_x}$ is defined **then** $\mathcal{D}_x \leftarrow \bot$

5:         **else**

6:             $\mathcal{D}_x \leftarrow (A \mapsto \emptyset)_{A \in N}$

            alt: $\mathcal{W}_v \leftarrow (A \mapsto 0)_{A \in N}$

7:     **while** $\mathcal{D}_{\bullet_g} = \bot$ **do**

8:         let $x \in V_g \cup E_g$ with $\mathcal{D}_x = \bot$ and $\mathcal{D}_y \neq \bot$ for all $y \in (V_{g{\downarrow_x}} \cup E_{g{\downarrow_x}}) \setminus ([\mathrm{ext}_{g{\downarrow_x}}] \cup \{x\})$

9:         **if** $x \in V_g$ **then** PARSE$_V(x)$

10:        **else** PARSE$_E(x)$

11:     **return** $\mathcal{D}_{\bullet_g}(S)$

    alt: **return** $\mathcal{W}_{\bullet_g}(S)$

Algorithm 3 Computing Derivations Trees of $g{\downarrow_v}$ for nodes $v \in V_g$

1: **function** PARSE$_V$(node v such that $\mathcal{D}_e(A) \neq \bot$ for all $e \in E_g$ with $\mathrm{src}_g(e) = v$)

2:     **if** v has out-degree 0 **then**

3:         $\mathcal{D}_v \leftarrow (A \mapsto \emptyset)_{A \in N}$

        alt: $\mathcal{W}_v \leftarrow (A \mapsto 0)_{A \in N}$

4:     **else**

5:         initialize $h = (V, E, \mathrm{att}, \mathrm{lab}, \mathrm{ext})$ as the following shallow graph:

6:         $E = \{e \in E_g \mid \mathrm{src}_g(e) = v\}$

7:         $V = \{v\} \cup \bigcup_{e \in E} \mathrm{reent}_g(e)$

8:         $\mathrm{ext} = \mathrm{ext}_{g{\downarrow_v}}$

9:         $\mathrm{att}(e) = vw$, where w is $\mathrm{reent}_g(e)$ ordered by $\preceq_g$, for each $e \in E$

10:       $\mathcal{D}_v \leftarrow$ SHALLOWPARSE($\{r \in R \mid r$ a duplication rule$\}$, h)

       alt: $\mathcal{W}_v \leftarrow$ SHALLOWPARSE($\{r \in R \mid r$ a duplication rule$\}$, h)

Algorithm 4 Computing Derivations Trees of $g{\downarrow}_e$ for edges $e \in E_g$

1: **function** PARSE_{E}(edge e s.t. $\mathcal{D}_y \neq \bot$ for all $y \in (V_{g(x)} \cup E_{g(x)}) \setminus ([\text{ext}_{g(x)}] \cup \{x\}))$

2:      $\mathcal{D}_e(A) \leftarrow \emptyset$ for all $A \in N$

      alt: $\mathcal{W}_e(A) \leftarrow 0$ for all $A \in N$

3:      **for** each deep rule $r = (A \to f)$ of arity ℓ **do**

4:          $\phi \leftarrow \text{MATCHING}(f, e)$

5:          **if** $\phi \neq$ null **then**

6:             $\mathcal{D}_e(A) \leftarrow \mathcal{D}_e(A) \cup r[\mathcal{D}_{\phi(\text{src}_f(e_1))}(\text{lab}_f(e_1)), \ldots, \mathcal{D}_{\phi(\text{src}_f(e_\ell))}(\text{lab}_f(e_\ell))]$

            alt: $\mathcal{W}_e(A) \leftarrow \mathcal{W}_e(A) + \omega(r) \cdot \prod_{i\in[\ell]} \mathcal{W}_{\phi(\text{src}_f(e_i))}(\text{lab}_f(e_i))\}$

---

gorithm 4 is described by Björklund et al. (2018) (using slightly different notation). It is based on the fact that, if $g{\downarrow}_e$ can be derived from a deep right-hand side f, then the mapping ϕ of the nodes in f to their images in $g{\downarrow}_e$ is uniquely determined by f and the reentrancies in $g{\downarrow}_e$, due to reentrancy and order preservation. As proved by Björklund et al. (2018), this makes it furthermore possible to compute $\phi = \text{MATCHING}(f, e)$ in linear time.

As mentioned above, the correctness of the computation of the sets $\mathcal{D}_x(A)$ was essentially shown by Björklund et al. (2018), and so we take it for granted here and use that fact to show inductively that the second version of the algorithm correctly computes the weights. Below, we assume for the sake of technical simplicity that the operations of the semiring $\mathcal{S}$ are computable in constant time. Clearly, the efficiency of the algorithm decreases accordingly if the operations a more complex. However, by the closedness of the class of polynomials under composition, the computation of weights stays polynomial whenever the operations of $\mathcal{S}$ are computable in polynomial time with respect to the input graph and the HRG.

Theorem 5.1. *Let $\prec$ be a suitable family of orders, and let η be a function mapping graphs to $\mathbb{N}$ such that both $\eta(g)$ and $\prec_g$ can be computed in time $\eta(g)$.[3] Then there is an algorithm which takes as input a graph g and an OPHG grammar $G = (\Sigma, N, S, R, \omega)$, and computes $\omega_G(g)$ in time $\mathcal{O}(\eta(g) + |g|^2 + |G|^2)$.*

Proof. With straightforward reformulations, the proof of the main theorem in (Björklund et al., 2018) shows that Algorithm 2 computes $\text{DT}_G(g)$ and runs in time $\mathcal{O}(\eta(g) + |g|^2 + |G|^2)$ if the time required for the explicit construction of deriva-

tion trees is neglected.[4] Together with the assumption that the operations of $\mathcal{S}$ can be computed in constant time, the latter means that the weight-computing version of Algorithm 2 runs in time $\mathcal{O}(\eta(g) + |g|^2 + |G|^2)$ as well. To complete the proof, it thus suffices to prove by induction that Algorithms 1–4 maintain the invariant that $\mathcal{W}_x(A) = \sum_{t\in\mathcal{D}_x(A)} \omega(t)$ for those edges and nodes x and those $A \in N$ such that $\mathcal{D}_x(A) \neq \bot$.

In the proof, for a set D of derivation trees, we abbreviate $\sum_{t\in D} \omega(t)$ by $\omega(D)$. We check the algorithms one by one. Note that the induction hypothesis states that the equation $\mathcal{W}_x(A) = \omega(\mathcal{D}_x(A))$ holds when the respective procedure is entered, and we have to show that it still holds afterwards. We use the fact that, by distributivity, for every rule $r = (A \to f)$ of arity ℓ and all sets $D_1, \ldots, D_\ell$ of derivation trees, it holds that

$$\omega(r(D_1, \ldots, D_\ell)) = \omega(r) \cdot \prod_{i\in[\ell]} \omega(D_i). \quad (1)$$

Procedure SHALLOWPARSE: We have to show that the two lines in the body of the loop starting in line 7 maintain the invariant. These lines change only $\mathcal{D}_e(A)$ and $\mathcal{W}_e(A)$, and after those two lines we have, for a rule $r = (A \to B^{\bullet\bullet})$ that fits Sib

$$\mathcal{W}_e(A) = \sum \omega(r^\ell) \cdot \prod_{i\in[\ell+1]} \mathcal{W}_{s_i}(B)$$

$$= \sum \omega(r^\ell) \cdot \prod_{i\in[\ell+1]} \omega(\mathcal{D}_{s_i}(B))$$

$$= \sum \omega(r^\ell\langle \mathcal{D}_{s_1}(B), \ldots, \mathcal{D}_{s_{\ell+1}}(B)\rangle)$$

$$= \omega(\mathcal{D}_e(A)).$$

Procedure PARSE: Only line 6 affects some $\mathcal{D}_x(A)$ and $\mathcal{W}_x(A)$. These lines obviously preserve the invariant.

---

[3]The function η describes the complexity of computing $\prec_g$, and the condition that it can be executed in time $\eta(g)$ corresponds to the usual requirement of time constructibility.

[4]Instead of computing the sets $\mathcal{D}_x(A)$, the algorithm in (Björklund et al., 2018) only computes, for every $x \in V_g \cup E_g$, the set of all $A \in N$ such that $\mathcal{D}_x(A) \neq \emptyset$.

Procedure PARSE$_\text{V}$: As before, line 3 respects the invariant. Concerning line 10, note that the two versions of SHALLOWPARSE return $(A \mapsto \mathcal{D}_e(A))_{A \in N}$ and $(A \mapsto \mathcal{W}_e(A))_{A \in N}$, respectively, for some edge e. By induction hypothesis, $\mathcal{W}_e(A) = \omega(\mathcal{D}_e(A))$ for all $A \in N$, which completes the argument.

Procedure PARSE$_\text{E}$: Once more, line 2 respects the invariant. Furthermore, if $D = \mathcal{D}_e(A)$ and $W = \mathcal{W}_e(A) = \omega(\mathcal{D}_e(A))$ before an execution of line 6 then, after this line,

$$\mathcal{W}_e(A) = W + \omega(r) \cdot \prod_{i \in [\ell]} \mathcal{W}_{\phi(\mathrm{src}_f(e_i))}(\mathrm{lab}_f(e_i))\}$$

$$= \omega(D) + \omega(r) \cdot \prod_{i \in [\ell]} \omega(\mathcal{D}_{\phi(\mathrm{src}_f(e_i))}(\mathrm{lab}_f(e_i)))$$

$$= \omega(D) + \omega(r[\mathcal{D}_{\phi(\mathrm{src}_f(e_1))}(\mathrm{lab}_f(e_1)),$$
$$\vdots$$
$$\mathcal{D}_{\phi(\mathrm{src}_f(e_\ell))}(\mathrm{lab}_f(e_\ell))]$$

$$= \omega(\mathcal{D}_e(A)).$$

This completes the correctness proof of the theorem. $\square$

As indicated before, it is worthwhile noticing that the first version of the parsing algorithm computes the set $\mathrm{DT}_G(g)$ in time $\mathcal{O}(\eta(g) + |g|^2 + |G|^2)$ if the sets $\mathcal{D}_x(A)$ are represented in a compact way as packed forests. This may be useful for further applications.

6 Conclusions

Semantic parsing is a necessary tool for the improvement of any number of natural language processing tools and the use of graphs as semantic models is becoming a standard approach. Abstract Meaning Representation is one example. There is, however no formal standard, and the algorithmic issues involved are largely unexplored. In particular, there are hardly any models for the formal description of weighted semantic graphs, despite the importance of probabilities and other kinds of weights in natural language processing for, e.g., resolving ambituities. In this contribution, we have taken a step towards resolving this situation by showing that order-preserving hyperedge replacement grammars can be extended with weights, without signficantly affecting the complexity of analysing a graph with respect to the grammar. We thus hope to have provided a useful building block for making semantic parsing practical.

To allow for efficient parsing, order-preserving hyperedge replacement grammars allow only for restricted forms of rules. In particular, the only way to create nodes of unlimited out-degree is to use so-called clone rules. Since clone rules are associative and commutative, we have opted to view the corresponding sections of the resulting derivation trees as unordered nodes of the appropriate degree and define the weight of these substructures as $w^\ell \prod_{i=0}^{\ell} w_i$, where w is the weight of the cloning rule (which is applied ℓ times) and $w_0, \dots, w_\ell$ are the weights of the subderivations. It may be worthwhile noting that, in cases where this is too restrictive, one may use a commutative product valuation monoid (Droste and Meinecke, 2010) as a weight structure. Such a valuation monoid comes with an additional *valuation function val* which takes an arbitrary multiset of weights to a generalized product. Then the expression above may be generalized to $w^\ell \cdot val(w_0, \dots, w_\ell)$ without making parsing more difficult.

References

Laura Banarescu, Claire Bonial, Shu Cai, Madalina Georgescu, Kira Griffitt, Ulf Hermjakob, Kevin Knight, Philipp Koehn, Martha Palmer, and Nathan Schneider. 2013. Abstract meaning representation for sembanking. In *Proc. 7th Linguistic Annotation Workshop, ACL 2013 Workshop*.

Michel Bauderon and Bruno Courcelle. 1987. Graph expressions and graph rewriting. *Mathematical Systems Theory*, 20:83–127.

Henrik Björklund, Frank Drewes, and Petter Ericson. 2016. Between a rock and a hard place – uniform parsing for hyperedge replacement DAG grammars. In *Proc. 10th Intl. Conf. on Language and Automata Theory and Applications*, volume 9618 of *Lecture Notes in Computer Science*, pages 521–532.

Henrik Björklund, Frank Drewes, Petter Ericson, and Florian Starke. 2018. Uniform parsing for hyperedge replacement grammars. Technical Report UMINF 18.13, Umeå University, http://www8.cs.umu.se/research/uminf/index.cgi. Submitted for publication.

Henrik Björklund, Johanna Björklund, and Petter Ericson. 2017. On the regularity and learnability of ordered DAG languages. In *Proc. 22nd International Conference on the Implementation and Application of Automata (CIAA'17)*, volume 10329 of *Lecture Notes in Computer Science*, pages 27–39. Springer.

David Chiang, Jacob Andreas, Daniel Bauer, Karl Moritz Hermann, Bevan Jones, and Kevin Knight. 2013. Parsing graphs with hyperedge replacement grammars. In *Proc. 51st Annual Meeting of the Association for Computational Linguistics (ACL 2013), Volume 1: Long Papers*, pages 924–932. The Association for Computer Linguistics.

David Chiang, Frank Drewes, Daniel Gildea, Adam Lopez, and Giorgio Satta. 2018. Weighted DAG automata for semantic graphs. *Computational Linguistics*, 44:119–186.

Bruno Courcelle. 1991a. The monadic second-order logic of graphs V: on closing the gap between definability and recognizability. *Theoretical Computer Science*, 80:153–202.

Bruno Courcelle. 1991b. The monadic second-order logic of graphs V: On closing the gap between definability and recognizability. *Theoretical Computer Science*, 80(2):153–202.

Bruno Courcelle and Joost Engelfriet. 2012. *Graph Structure and Monadic Second-Order Logic – A Language-Theoretic Approach*. Cambridge University Press.

Frank Drewes, Annegret Habel, and Hans-Jörg Kreowski. 1997. Hyperedge replacement graph grammars. In G. Rozenberg, editor, *Handbook of Graph Grammars and Computing by Graph Transformation. Vol. 1: Foundations*, chapter 2, pages 95–162. World Scientific.

Frank Drewes, Berthold Hoffmann, and Mark Minas. 2015. Predictive top-down parsing for hyperedge replacement grammars. In *Proc. 8th Intl. Conf. on Graph Transformation (ICGT'15)*, Lecture Notes in Computer Science.

Frank Drewes, Berthold Hoffmann, and Mark Minas. 2017. Predictive shift-reduce parsing for hyperedge replacement grammars. In *Proc. 10th Intl. Conf. on Graph Transformation (ICGT'17)*, volume 10373 of Lecture Notes in Computer Science, pages 106–122.

Manfred Droste and Ingmar Meinecke. 2010. Describing average- and longtime-behavior by weighted mso logics. In *Proc. 35th Intl. Symp. on Mathematical Foundations of Computer Science (MFCS 2010)*, volume 6281 of *Lecture Notes in Computer Science*, pages 537–548.

Sorcha Gilroy, Adam Lopez, and Sebastian Maneth. 2017. Parsing graphs with regular graph grammars. In *Proc. 6th Joint Conf. on Lexical and Computational Semantics (*SEM 2017)*, pages 199–208.

Joshua Goodman. 1999. Semiring parsing. *Computational Linguistics*, 25:573–605.

Jonas Groschwitz, Alexander Koller, and Christoph Teichmann. 2015. Graph parsing with s-graph grammars. In *Proc. 53rd Ann. Meeting of the Association for Computational Linguistics and the 7th Intl. Joint Conf. on Natural Language Processing (Volume 1: Long Papers)*, pages 1481–1490.

Annegret Habel. 1992. *Hyperedge Replacement: Grammars and Languages*, volume 643 of *Lecture Notes in Computer Science*. Springer.

Annegret Habel and Hans-Jörg Kreowski. 1987. May we introduce to you: Hyperedge replacement. In *Proceedings of the Third Intl. Workshop on Graph Grammars and Their Application to Computer Science*, volume 291 of Lecture Notes in Computer Science, pages 15–26. Springer.

Bevan Jones, Jacob Andreas, Daniel Bauer, Karl Moritz Hermann, and Kevin Knight. 2012. Semantics-based machine translation with hyperedge replacement grammars. In *Proc. 24th Intl. Conf. on Computational Linguistics (COLING 2012): Technical Papers*, pages 1359–1376.

Alexander Koller. 2015. Semantic construction with graph grammars. In *Proc. 11th Intl. Conf. on Computational Semantics*, pages 228–238.

Sensing Tree Automata as a Model of Syntactic Dependencies

Thomas Graf
Department of Linguistics
Stony Brook University
Stony Brook, NY, USA
mail@thomasgraf.net

Aniello De Santo
Department of Linguistics
Stony Brook University
Stony Brook, NY, USA
aniello.desanto@stonybrook.edu

Abstract

Various aspects of syntax have recently been characterized in subregular terms. However, these characterizations operate over very different representations, including string encodings of c-command relations as well as tiers projected from derivation trees. We present a way to unify these approaches via sensing tree automata over Minimalist grammar dependency trees. Sensing tree automata are deterministic top-down tree automata that may inspect the labels of all daughter nodes before assigning them specific states. It is already known that these automata cannot correctly enforce all movement dependencies in Minimalist grammars, but we show that this result no longer holds if one takes into account several well-established empirical restrictions on movement. Sensing tree automata thus furnish a strong yet uniform upper bound on the complexity of syntactic dependencies.

1 Introduction

This paper proposes a novel, unified upper bound on the subregular complexity of syntactic dependencies. Since the merit of this result might be opaque to a reader who is not intimately familiar with the most recent developments in subregular complexity, we start out with a very detailed background discussion.

The subregular program seeks to identify formal machinery that provides a tighter characterization of natural language than the familiar classes of the Chomsky hierarchy. Subregular phonology took as its vantage point the well-known result that phonology is regular (Johnson, 1972; Kaplan and Kay, 1994) and then identified proper subclasses that are still powerful enough for specific types of phonological dependencies. Among these subclasses are SL, SP (Rogers et al., 2010), IBSP (Graf, 2017, 2018a), TSL (Heinz et al., 2011; McMullin, 2016), and several extensions of the latter

(Baek, 2018; Graf and Mayer, 2018). The highly restrictive nature of these classes has allowed for new learning algorithms (Heinz et al., 2012; Jardine and McMullin, 2017) and also furnishes computational explanations for typological gaps.

Syntax cannot be subregular in this strict sense by virtue of being at least mildly context-sensitive (Huybregts, 1984; Shieber, 1985; Michaelis and Kracht, 1997; Kobele, 2006, a.o.). However, linguists formulate syntactic dependencies over trees, not strings. This shift in perspective has also taken place within formal grammar, first with model-theoretic syntax (Blackburn et al. 1993; Backofen et al. 1995; Cornell and Rogers 1998; Rogers 1998, 2003, a.o.) and then the two-step approach (see Morawietz 2003, Mönnich 2006, and references therein). The two-step approach highlighted how the mildly context-sensitive nature of syntax arises from the interaction of two finite-state components: a regular tree language that encodes a kind of "deep structure" and a finite-state tree transduction to the intended "surface structure". This perspective has proven particularly fruitful in Minimalist grammars (MGs; Stabler, 1997, 2011a), where derivation trees provide the regular tree language and the transduction is a formal analogue to Chomsky's notion of movement (Kobele et al., 2007; Graf, 2012b). With the regular nature of syntax properly identified, a subregular characterization of syntax is suddenly feasible.

Graf (2012a) provided the first subregular (un)definability results for MG derivation tree languages, but these results were expanded on only recently. Notably, all follow-up work strived to remain closer to the classes that enjoy prominence in subregular phonology. This, however, also led to a marked divergence in approaches. Graf (2018b) operates directly over MG derivation trees. Following the tradition of model-theoretic syntax, Graf equates the MG operations Merge and Move with constraints on MG derivation tree languages and shows that they belong to the tree analogue of the subregular string class TSL. This view is also adopted by Vu (2018) and Vu et al. (2019) in the analysis of negative-polarity items and case licensing, respectively. Graf and Shafiei (2019) and Shafiei and Graf (2019), on the other hand, pursue a purely string-based perspective of syntactic dependencies. For each node they identify its string of c-commanders, the shape of which must follow the constraints imposed by, say, Principle A or NPI-licensing. When construed as such string constraints, syntactic licensing conditions not only turn out to be subregular, they also fit into classes that have been proposed for subregular phonology. The work so far thus has unearthed two distinct subregularity results: MG operations are subregular over MG derivation trees, and licensing conditions are subregular over a specific string representation grounded in c-command (cf. Frank and Vijay-Shanker, 2001).

Even though each perspective is worthwhile and has proven very fruitful, their apparent incommensurability raises the question how these two notions of subregularity can be brought to bear on each other. The central contribution of our paper is a uniform upper bound on syntax that encompasses both MG operations and licensing conditions. This upper bound takes the form of sensing tree automata (STAs) operating over dependency tree representations of MG derivations (these dependency trees are distinct from the MG dependency trees of Boston et al. 2010). STAs provide a minimal amount of look-ahead to deterministic top-down tree automata: the automaton may inspect the labels of all daughter nodes before assigning them specific states. Far more than just a mathematical curiosity, limiting syntax to STA-recognizable constraints over MG dependency trees is very natural in several respects:

1. MG dependency trees are a natural encoding of head-argument relations.

2. STAs explain why licensing conditions are mediated by c-command instead of the many alternative command relations one could imagine (Barker and Pullum, 1990).

3. In order for movement to be regulated by STAs, it must obey additional restrictions beyond those of the standard MG formalism. These restrictions coincide with well-known empirical phenomena such as the Specifier Island Constraint and the Coordinate Structure Constraint.

4. As a single STA can handle movement and licensing conditions at the same time, it is unsurprising that the two occasionally interact, e.g. when movement induces a licensing configuration or violates one.

5. Since STAs are deterministic top-down automata with a minimal amount of look-ahead, they are a natural match for top-down parsing, which has been argued to play a central role in human sentence processing (Stabler, 2013; Kobele et al., 2013; Graf et al., 2017).

Hence our contribution does not merely unify the two existing approaches to subregular syntax, it also accounts for empirical aspects of syntax that the latter leave unexplained. At the same time, we do not intend for our perspective to supplant the existing ones. Each one provides useful insights and can now be safely pursued in the knowledge that there is a principled formal connection to the other approaches.

The paper is laid out as follows: Section 2 introduces our mathematical notation for trees and tree languages (§2.1), which form the basis for our definition of STAs (§2.2). We then define MGs and their dependency trees in §3. In order to simplify some of the subsequent proofs, we do not follow the standard definitions and instead adopt a format that is partly inspired by Kobele et al. (2007). Sections 4 and 5 cover the two core results of this paper: dependency trees of MGs with the Specifier Island Constraint are STA-recognizable, and so are licensing conditions based on c-command. The last section briefly sketches some extensions of these results, including interactions of movement and c-command (§6.1), subcommand (§6.2), recognizability of MG derivation trees (§6.3), adjunct islands and the coordinate structure constraint (§6.4, §6.5), and connections to top-down parsing (§6.6).

2 Preliminaries

2.1 Trees and Tree Languages

A *ranked alphabet* Σ is a finite set of symbols, each one of which has a *rank* or *arity* assigned by the function $r : \Sigma \to \mathbb{N}$. We write $\Sigma^{(n)}$ to denote $\{\sigma \in \Sigma \mid r(\sigma) = n\}$, and $\sigma^{(n)}$ indicates that

σ has rank n. Given a ranked alphabet Σ, the set $T(\Sigma)$ of Σ-*trees* contains all $\sigma^{(0)}$ and all terms $\sigma^{(n)}(t_1, \ldots, t_n)$ $(n \geq 0)$ such that $t_1, \ldots, t_n \in T(\Sigma)$.

Example 1. Given $\Sigma := \{a^{(0)}, b^{(0)}, c^{(2)}, d^{(2)}\}$, $T(\Sigma)$ is an infinite sets that contains, among others, the term $d(c(b, b), d(b, a))$. This term corresponds to the tree below:

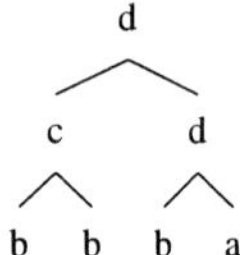

If Σ consisted only of $c^{(2)}$ and $d^{(2)}$, then $T(\Sigma)$ would be empty. ⌋

Given a term $m^{(n)}(s_1, \ldots, s_n)$ where each s_i is a subtree with root d_i, we call m the *mother* of the *daughters* $d_1, \ldots, d_n$ $(1 \leq i \leq n)$. If two distinct nodes have the same mother, they are *siblings*. We use the term *proper dominance* for the transitive closure of the mother-of relation, and *reflexive dominance* for the reflexive, transitive closure. A node a is an *ancestor* of node n iff a properly dominates n.

Every node u in a Σ-tree has a unique address $g(u)$ as defined in Gorn (1967). The subtree of Σ-tree t rooted in Gorn address $g(u)$ is denoted by $t/g(u)$. Given two Σ-trees s and t and node u in t, $t[g(u) \leftarrow s]$ is the result of replacing $t/g(u)$ in t by s. Throughout the paper, we write u instead of $g(u)$ so that t/u and $t[u \leftarrow s]$ are shorthands for $t/g(u)$ and $t[g(u) \leftarrow s]$, respectively.

2.2 Sensing Tree Automata

A *sensing tree automaton* (STA; Martens et al., 2008) is a deterministic top-down tree automaton that may also take the labels of a node's daughters into account before assigning states to them. In other words, these automata have a finite look-ahead of 1. This intuition is reflected in our definition of *sensing rules* below. Following the format of Comon et al. (2008, p. 38) for top-down tree automata, we deliberately define sensing rules in a tree transducer format. This is a marked deviation from Martens et al. (2008), but it should make it easier for future work to extend our perspective from constraints to transformations.

Let Σ be some ranked alphabet and Q a set of symbols of rank 1 disjoint from Σ. Members of Q are called *states*. In addition, X is a countably infinite set of variable symbols $(X \cap \Sigma = X \cap$

$Q = \emptyset)$. For the rest of this section, we use $\vec{\sigma_i}$ as a shorthand for $\sigma_i(x_{i_1}, \ldots, x_{i_n})$, with $\sigma_i \in \Sigma^{(n)}$ and $x_{i_1}, \ldots, x_{i_n} \in X$.

Definition 1 (Sensing rule). A *sensing rule* over alphabet Σ and state set Q is an expression of the form $q(\sigma^{(n)}(\vec{\sigma_1}, \ldots, \vec{\sigma_n})) \rightarrow \sigma^{(n)}(q_1(\vec{\sigma_1}), \ldots, q_n(\vec{\sigma_n}))$ such that $\sigma, \sigma_1, \ldots, \sigma_n \in \Sigma$ and $q, q_1, \ldots, q_n \in Q$. ⌋

Note that for $\sigma^{(0)}$, sensing rules are of the form $q(\sigma^{(0)}) \rightarrow \sigma^{(0)}$. Such sensing rules effectively remove states from leaf nodes, whereas sensing rules for $\sigma^{(\geq 1)}$ remove the state of σ and instead add a state on top of each daughter of σ.

Definition 2 (Sensing relation). Let t and t' be in $T(\Sigma \cup Q)$, and suppose δ is some sensing rule over Σ and Q that has the form $q(\sigma^{(n)}(\vec{\sigma_1}, \ldots, \vec{\sigma_n})) \rightarrow \sigma^{(n)}(q_1(\vec{\sigma_1}), \ldots, q_n(\vec{\sigma_n}))$. Then the relation $\rightarrow_\delta$ holds between t and t' ($t \rightarrow_\delta t'$) iff t contains a subtree $s := q(\sigma^{(n)}(s_1, \ldots, s_n))$ such that each s_i $(1 \leq i \leq n)$ is a Σ-tree with root σ_i, and t' is the result of replacing s in t with $\sigma^{(n)}(q_1(s_1), \ldots, q_n(s_n))$.

Given a set Δ of sensing rules, we write $t \rightarrow_\Delta t'$ iff $t \rightarrow_\delta t'$ for some $\delta \in \Delta$. The reflexive, transitive closure of $\rightarrow_\Delta$ is denoted by $\rightarrow_\Delta^*$. ⌋

Intuitively, $t \rightarrow_\Delta^* t'$ iff t' can be obtained from t by a sequence of sensing rules.

Definition 3 (Sensing tree automaton). A *sensing tree automaton* (STA) over Σ is a 4-tuple $A := \langle Q, \Sigma, q_I, \Delta \rangle$ such that Q is a finite set of states disjoint from alphabet Σ, $q_I \in Q$ is the initial state, and Δ is a finite set of *sensing rules* over Σ and Q. The tree language recognized by A is $L(A) := \{t \in T(\Sigma) \mid q_I(t) \rightarrow_\Delta^* t\}$. The class of all tree languages that are recognized by at least one STA is called STA. ⌋

An STA thus recognizes a tree t iff there is a sequence of sensing rules that starts from the initial state q_I and passes states through t until they are all removed again at the leaves of t. If the STA ever gets stuck because there is no suitable sensing rule, t is rejected.

Example 2. Suppose $\Sigma := \{a^{(0)}, b^{(0)}, c^{(2)}, d^{(2)}\}$, and consider the STA with $Q := \{0, 1\}$, where 0 is the initial state. The automaton uses all sensing rules of the following form:

- $1(\sigma(\vec{\sigma_1}, \vec{\sigma_2})) \rightarrow \sigma(1(\vec{\sigma_1}), 1(\vec{\sigma_2}))$,
  for $\sigma \in \{c, d\}$ and $\sigma_1, \sigma_2 \in \{a, b, c, d\}$,

14

- $0(\sigma(\vec{\sigma_1}, \vec{\sigma_2})) \rightarrow \sigma(0(\vec{\sigma_1}), 0(\vec{\sigma_2}))$,
  for $\sigma \in \{c, d\}$ and $\sigma_1, \sigma_2 \in \{a, b, d\}$,

- $0(\sigma(\vec{\sigma_1}, \vec{\sigma_2})) \rightarrow \sigma(q_0(\vec{\sigma_1}), q_1(\vec{\sigma_2}))$,
  for $\sigma \in \{c, d\}$, $\sigma_1, \sigma_2 \in \{a, b, c, d\}$, and
  $q_i := 1$ iff $\sigma_{2-i} := c$ (where $i \in \{0, 1\}$),

- $q(b) \rightarrow b$,
  for $q \in \{0, 1\}$,

- $1(a) \rightarrow a$.

The STA only accepts those Σ-trees where each a is c-commanded by some c (that is to say, c must be the sibling of an ancestor of a). It thus emulates a simplified version of Principle A. ⌐

It will also be convenient in some proofs to employ a substitution-based characterization of the tree languages that are recognized by STAs. The characterization capitalizes on the fact that every STA only has a look-ahead of 1. Consequently, the state it assigns to a node n depends only on three components: I) the label of n and its siblings, II) the states assigned to n's ancestors, and III) the states assigned to the siblings of each ancestor. If all of those are kept constant, n will always receive the same state no matter what the rest of the tree looks like.

Definition 4 (Spine closure). Given a node u of some Σ-tree t, $\mathrm{lsib}^t(u)$ is the string consisting of the label of u's left siblings (if they exist) followed by the label of u. Analogously, $\mathrm{rsib}^t(u)$ is the string consisting of the label of u and the label of its right siblings (if they exist). Also, $\bigtriangledown$ and $\Downarrow$ are two distinguished symbols not in Σ.[1] Let $u_1, \ldots, u_n$ be the shortest path of nodes extending from the root of t to u. That is to say, each u_i is the mother of u_{i+1} $(1 \le i < n)$, u_1 is the root, and $u_n = u$. By $\mathrm{spine}^t(u)$ we denote the string recursively defined by $\mathrm{spine}^t(u_1) = u_1 \bigtriangledown u_1$ and $\mathrm{spine}^t(u_1, \ldots, u_n) = \mathrm{spine}^t(u_1, \ldots, u_{n-1}) \Downarrow \mathrm{lsib}^t(u_n) \bigtriangledown \mathrm{rsib}^t(u_n)$ A regular tree language L is *spine-closed* iff it holds for all trees $s, t \in L$ and nodes u and v belonging to s and t, respectively, that $\mathrm{spine}^s(u) = \mathrm{spine}^t(v)$ implies $s[u \leftarrow t/v] \in L$. ⌐

Theorem 1 (Martens 2006). *A regular tree language L belongs to the class* STA *iff L is spine-closed.*

---

[1] Martens et al. (2008) use # instead of $\Downarrow$. We prefer the latter as it emphasizes visually that this symbol marks the start of a string at the next lower level in the tree.

Example 3. Consider the left tree l and right tree r in Fig. 1. Let $l0$ and $r0$ be the left daughter of the root in l and r, respectively. Then $\mathrm{spine}^l(l0) = \mathrm{spine}^r(r0) = merge \bigtriangledown merge \Downarrow merge \bigtriangledown merge\ merge$. Hence a tree language that contains both l and r is an STA language iff it also contains $r[r0 \leftarrow r/l0]$. ⌐

In linguistic terms, spine-closure tells us that two subtrees s/u and t/v with identical root labels can be freely exchanged whenever they have the same ancestors and the same c-commanders. This will be of great importance throughout this paper.

3 Minimalist Grammars

Minimalist grammars (MGs; Stabler, 1997) are a formalization of Minimalist syntax (Chomsky, 1995). Readers who are unfamiliar with the formalism should consult Stabler (2011a) for a more accessible introduction.

Every MG consists of a finite set of feature annotated lexical items. Each lexical item is a pair of a phonetic exponent and a finite, non-empty string of features. There are four types of features, whose job it is to trigger the structure-building operations *Merge* and *Move*. Merge establishes head-argument relations and is triggered when a *selector feature* F^+ on a head finds a matching *category feature* F^- on an argument. For example, the noun *guest* carries a feature N^-, for which we also write guest :: N^-. It can be merged with the :: $\mathrm{N}^+\mathrm{D}^-$ to yield a DP, thanks to the matching category and selector features. Move displaces a subtree from its current position to a higher position in the syntactic structure. Move takes place when a *licensor features* f^+ on a head that provides a landing site can be checked by a corresponding *licensee feature* f^- on a mover. The order of features on a lexical item determines the order in which the corresponding operations are triggered. Hence the determiner which :: $\mathrm{N}^+\mathrm{D}^-\mathrm{wh}^-$ would first select a noun phrase, get merged with a head looking for a DP, and then undergo wh-movement.

The sequence of Merge and Move steps is commonly represented as a derivation tree, e.g. in Fig. 2. However, we will use a dependency tree representation instead. Our dependency trees are merely a more compact encoding of MG derivation trees and have no connection to the MG dependency trees of Boston et al. (2010). We will

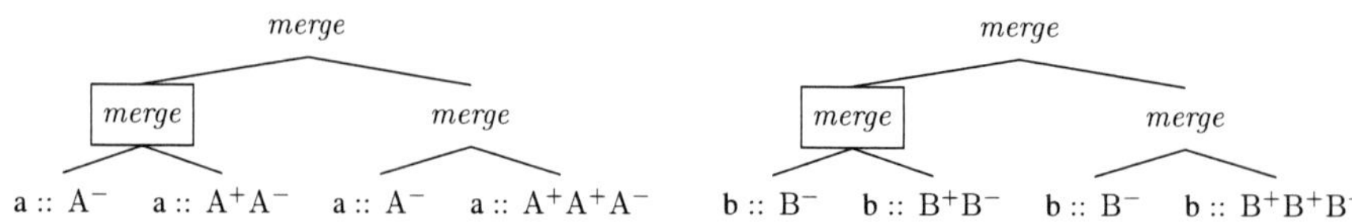

Figure 1: MG derivation tree languages are not recognizable by STAs because they are not spine-closed.

directly define MGs as sets of well-formed dependency trees, mirroring earlier definitions in terms of derivation trees (primarily Kobele et al. 2007).

We pick a ranked alphabet Σ such that $\Sigma^{(n)}$ contains all lexical items of MG G, and only those, that carry exactly n selector features. For simplicity, we use G to also refer to this alphabet. Not every G-tree is a well-formed dependency tree, though, due to the constraints of the MG feature calculus. The calculus is illustrated in Fig. 2, where each node in the dependency tree is annotated with the feature configuration corresponding to its subtree. We formalize this calculus via a recursive function feat that computes these values based on more primitive functions for the feature checking steps that trigger Merge and Move.

Definition 5 (Dependency tree language). If G is an MG with n distinct licensee features, then the set $\mathrm{dep}(G)$ of well-formed dependency trees of G is $\{t \in T(G) \mid \mathrm{feat}(t) = \langle \mathrm{C}^-, \varepsilon_1, \ldots, \varepsilon_n \rangle\}$. ⌟

The remainder of this section defines feat in terms of feature checking operators M and $\otimes$ for Move and Merge, respectively. All operations manipulate one or more sequences of feature strings. The Shortest Move Constraint (SMC) will be used to filter out illicit sequences.

Definition 6 (SMC). Let G be an MG and Lce its set of n licensee features. Given a sequence $s := s_1, \ldots, s_m$ ($m \geq 0$) of strings in Lce*, SMC : $(\mathrm{Lce}^*)^* \to (\mathrm{Lce}^*)^*$ is undefined for s if there are s_i and s_j ($1 \leq i \neq j \leq m$) that start with the same licensee feature. Otherwise, SMC maps s to s itself. ⌟

The SMC ensures that Move is unambiguous in the sense that there can never be more than one active mover of a specific type (wh, topicalization, and so on). It is an integral part of MGs, and removing it would greatly alter their expressivity (Salvati, 2011).

Next we add a helper function sort that orders SMC-approved sequences based on the first licensee feature of each feature string.

Definition 7 (sort). Let G and Lce be as before. Now fix some bijection b between Lce and $\{1, \ldots, n\}$. Then sort maps $s := s_1, \ldots, s_m \in$ Lce* to the sequence $s'_1, \ldots, s'_n$ such that $s'_i := s_j$ if s_j starts with f and $b(f) = i$; otherwise, $s'_i := \varepsilon$ ($1 \leq i \leq n, 1 \leq j \leq m$). ⌟

Now we can finally define M for Move, which is also a crucial part of the Merge operator $\otimes$. Throughout we use γ as a shorthand for any string of features, and δ for a (possibly empty) string of licensee features.

Definition 8 (Move). Suppose that expression e is $\langle f\gamma, \delta_1, \ldots, \mathrm{f}^-\delta_i, \ldots, \delta_n \rangle$ for some licensee feature f^- and $1 \leq i \leq n$. Then

$$M(e) := M(\langle \gamma, \mathrm{sort}(\mathrm{SMC}(\delta_1, \ldots, \delta_i, \ldots, \delta_n)) \rangle)$$

if f is the licensor feature f^+, and e otherwise. ⌟

Definition 9 (Merge). Given two expressions $e := \langle f\gamma, \delta_1, \ldots, \delta_m \rangle$ and $e' := \langle f'\delta, \delta_{m+1}, \ldots, \delta_z \rangle$, $e \otimes e'$ is

$$M(\langle \gamma, \mathrm{sort}(\mathrm{SMC}(\delta, \delta_1, \ldots, \delta_m, \delta_{m+1}, \ldots, \delta_z)) \rangle)$$

if f is some selector feature F^+ and f' the matching category feature F^-. In all other cases, $\otimes$ is undefined. ⌟

Note how Merge is always followed by an application of the Move operator, but this does not trigger any feature checking unless γ starts with a licensor feature. Merge steps are thus interleaved with movement checks, not all of which may actually result in movement.

The operators M and $\otimes$ on their own do not narrow down the set of G-trees. They are invoked as part of a recursive function feat over G-trees that computes the feature values of subtrees, as already expressed in Def. 5.

Definition 10 (feat). The partial function feat recursively maps MG dependency trees to feature expressions. For lexical items, $\mathrm{feat}(\sigma :: \phi) := \langle \phi, \varepsilon_1, \ldots, \varepsilon_n \rangle$. If $t := \sigma(\sigma_1, \ldots, \sigma_z)$, then $\mathrm{feat}(t) := ((\mathrm{feat}(\sigma) \otimes \mathrm{feat}(\sigma_z)) \otimes \cdots) \otimes \mathrm{feat}(\sigma_1)$. ⌟

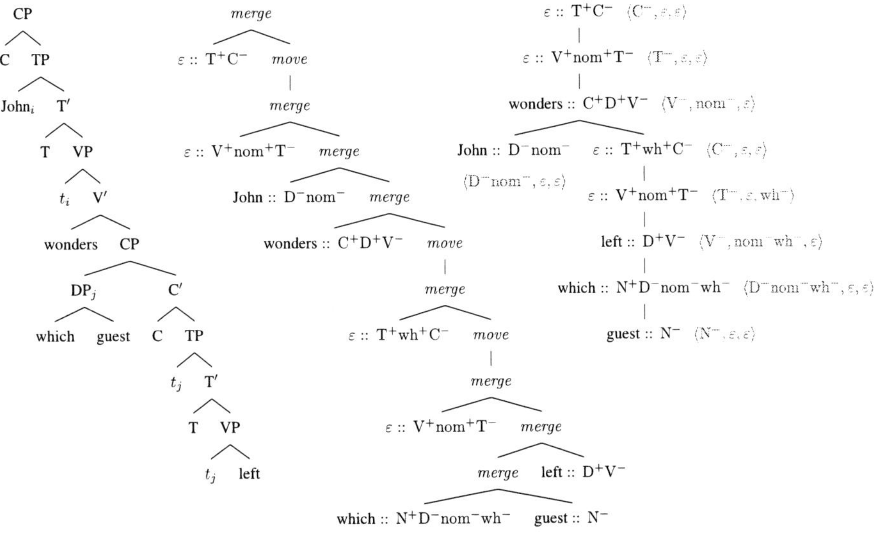

Figure 2: X′-bar tree, corresponding MG derivation tree, and (feat-annotated) MG dependency tree

Two important lemmata follow immediately from the preceding definitions.

Lemma 1. *Let G be an MG and s a G-tree. Then it holds for every $t \in \text{dep}(G)$ with node u that $t[u \leftarrow s] \in \text{dep}(G)$ iff $\text{feat}(s) = \text{feat}(t/u)$.*

Lemma 2. *Let G be an MG with n distinct licensee features and s a subtree of some $t \in \text{dep}(G)$. Then $\text{feat}(s)$ must be of the form $\langle F^- \delta, \delta_1, \ldots, \delta_n \rangle$ for some category feature F^-.*

Lemma 2 is apparent from the derivation tree example in Fig. 2. It is a minor extension of the well-known fact that a lexical item may occur in a well-formed MG derivation iff its feature string is of the form $\phi F \delta$, where ϕ is either ε or a selector feature followed by 0 or more selector and licensor features, F is a category feature, and δ is a (possibly empty) string of licensee features.

4 Merge and Move via STAs

With all the preliminaries in place, we can finally turn to the core results regarding the STA-recognizability of MGs with respect to Merge and Move. The next sections then extend this to licensing conditions and some other special cases.

Graf (2012a) uses the argument from example 3 in §2.2 to prove that MG derivation tree languages are not STA languages. However, this proof does not carry over to MG dependency trees. Adopting the terminology of Graf (2012a), we use MDEP[*merge, move*] for the full class of MG de-

pendency tree languages and MDEP[*merge*] for the subclass of movement-free MGs (no lexical item carries any licensee features).

Theorem 2. MDEP[*merge*] $\subsetneq$ STA

This is just a corollary of a more fundamental property of MGs. For any arbitrary $L \in$ MDEP[*merge*] and nodes u and v of $s, t \in L$, $\text{spine}^s(u) = \text{spine}^t(v)$ necessarily entails that $\text{feat}(s/u) = \text{feat}(t/v)$, so that Theorem 2 immediately follows from Lemma 1. We omit a full proof here as Lemma 3 will cover a more complex case that subsumes this one.

Even with the dependency tree format, though, STAs are too weak for standard MGs with both Merge and Move.

Theorem 3. MDEP[*merge, move*] *and* STA *are incomparable.*

Proof. Consider the dependency trees l and r in Fig. 3. The respective instances of the $::$ N^+D^- have different values under feat ($\langle D^-, \varepsilon \rangle$ and $\langle D^-, wh^- \rangle$, respectively). By Lemma 1, then, their subtrees are not interchangeable even though $\text{spine}^l(\text{the} :: N^+D^-) = \text{spine}^r(\text{the} :: N^+D^-)$. □

The example in Fig. 3 is peculiar, though. The left dependency tree encodes the derivation for *Who does a teacher of like the father of John*, which is severely degraded. It has been argued that such cases of left-branch subextraction are generally forbidden. MGs can be equipped with the *Specifier Island Constraint* (SpIC) to

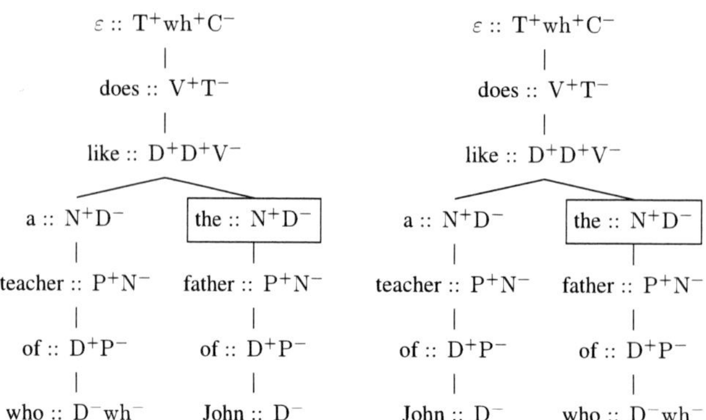

Figure 3: Even though the boxed nodes have the same spine, their subtrees cannot be exchanged.

rule out movement from within a specifier. This takes the form of an additional restriction on feat that only allows complements to properly contain unchecked licensee features. Since the complement of a head is its rightmost daughter in the dependency tree (rather than the leftmost one), the SpIC amounts to a restriction on all daughters except the last one.

Definition 11 (SpIC). Suppose $s_1, \ldots, s_m$ are G-trees and $\sigma \in G^{(m)}$. Then $\mathrm{feat}(\sigma(s_1, \ldots, s_m))$ is undefined if there is an $i < m$ and $1 \leq j \leq n$ such that $\mathrm{feat}(s_i)$ is of the form $\langle \gamma, \delta_1, \ldots, \delta_j, \ldots, \delta_n \rangle$ and $\delta_j \neq \varepsilon$.

Note that any licensee features on the head of a specifier are part of γ, not δ_j, so specifiers can still move without violating the SpIC. Only extraction of a proper subtree from within a specifier is not allowed. Also note that our version of the SpIC only bans extraction from base specifiers, but not from specifiers that are derived via movement. This is why it can be easily stated over dependency trees. Even so, this limited version of the SpIC greatly limits the weak generative capacity of MGs with the SMC, while still keeping them mildly context-sensitive (Michaelis, 2004, 2009; Kobele and Michaelis, 2011). For our purposes, though, the major contribution of the SpIC is that it also lowers the complexity of MG dependency trees into STA.

Lemma 3. *Given an MG G that obeys the SpIC, pick arbitrary nodes u and v of $s, t \in \mathrm{dep}(G)$, respectively. If $\mathrm{spine}^s(u) = \mathrm{spine}^t(v)$, then $\mathrm{feat}(s/u) = \mathrm{feat}(t/v)$.*

Proof. Since s and t are well-formed, both $\mathrm{feat}(s/u)$ and $\mathrm{feat}(t/v)$ must be defined. By Lemma 2 we may assume w.l.o.g. that $\mathrm{feat}(s/u) := \langle \mathrm{F}^-\delta, \delta_1, \ldots, \delta_n \rangle$ and $\mathrm{feat}(t/v) := \langle \mathrm{F}'^-\delta', \delta_1', \ldots, \delta_n' \rangle$. As u and v have identical spines, u and v themselves must be identical. Therefore both $\mathrm{F}^- = \mathrm{F}'^-$ and $\delta = \delta'$ hold. Now suppose that $\delta_i \neq \delta_i'$ for some $1 \leq i \leq n$. Then either s/u or t/v is missing a licensee feature f^- that is present in the other. Suppose it is s/u that is missing a feature present in t/v. Note that this immediately entails by the SpIC that t/v is a complement, wherefore s/u is also a complement because u and v have identical spines. Since both s and t are well-formed, whatever feature is missing in s/u must occur somewhere else in s to match some f^+ in the spine of u. But by the SpIC, f^- cannot occur properly inside any specifier. We already know that s/u is a complement, so f^- can only occur on an ancestor of u or on one of its left siblings. But then it would occur on a node in the spine of u, which contradicts our initial assumption that $\mathrm{spine}^s(u) = \mathrm{spine}^t(v)$. Hence $\delta_i = \delta_i'$ after all, wherefore $\mathrm{feat}(s/u) = \mathrm{feat}(t/v)$. $\square$

Theorem 4. *For every MG G that obeys the SpIC, it holds that $\mathrm{dep}(G) \in \mathrm{STA}$.*

Proof. Lemma 1 and 3 jointly imply that $\mathrm{dep}(G)$ is spine-closed, which guarantees that it can be recognized by an STA (Thm. 1). $\square$

Intuitively, Theorem 4 holds because the SpIC creates a unique "elsewhere case" for missing movers. Suppose that an STA is at node n in a G-tree t. Since it has processed t top-down, it knows exactly which movers it has to look for by virtue of the licensor features it has come across. With its look-ahead of 1, the STA can scan the daughters of n to see if any of them carry some of the desired licensee features. Any licensee features that are not among them must be embedded deeper in the tree. Due to the SpIC, though, they can only reside in the complement, i.e. the subtree rooted in the rightmost daughter of n.

The SpIC is just one way of creating such an elsewhere case. STA-recognizability would also hold if movers could only escape from the leftmost argument, or if the label of the selecting head decides which one of its arguments can be extracted from. Extraction from arguments could also be parameterized for each feature so that wh-movers may only leave complements whereas topicalization is only allowed from the last but one argument, if it exists. Or the STA could switch between these four constraints depending on the number of ancestors of the current node *modulo* 4. STA-recognizability holds as long as distributing the head's δ_i across its arguments is fully deterministic based on the information available to a sensing rule (current state, label of selecting head, labels of selected heads). Hence the class of STA-recognizable MG dependency tree languages is larger than what is allowed by the SpIC.

This does not change the fact, though, that the initial finding of Graf (2012a) regarding the insufficiency of STAs is incomplete. In the case of Merge, the insufficiency disappears with the more compact representation format of MG dependency trees. Alternatively, one could also keep the derivation tree format while increasing the STA look-ahead beyond just one level — this is a point we will revisit soon in §6.2 and §6.3, for very different reasons. With respect to Move, the choice of representation format is immaterial. Neither derivation trees nor dependency trees make movement as defined in MGs STA-recognizable. However, the SpIC does make movement STA-recognizable because specifiers do not need to be probed deeper than their head. For MG dependency trees, this coincides with the 1-level look-ahead of STAs, whereas derivation trees once again require a more generous look-ahead window. The choice of representation thus has an impact on the amount of required look-ahead, but the essence of our STA-recognizability result for MGs rests on the SpIC, not the tree format.

5 STAs and C-Command Conditions

The previous section has successfully established that the central operations of MGs can be han-dled by STAs, assuming that I) they are construed as constraints on MG dependency trees, and II) movement is subject to the SpIC. But the structure-building operations Merge and Move are just one part of syntax. Licensing conditions also play a major role, in particular those rooted in c-command. This section shows that these conditions are also captured by STAs.

Licensing conditions based on c-command are ubiquitous in the syntactic literature. They were recently studied from a subregular perspective by Graf and Shafiei (2019) and Shafiei and Graf (2019). Both papers use similar ideas, but define them very differently. We adopt the formalism of Graf and Shafiei (2019) because it defines all essential concepts directly in terms of dependency trees.

Definition 12 (C-string). Let t be some MG dependency tree. For every node n of t in configuration $m(d_1, \ldots, d_i, n, d_{i+1}, \ldots, d_j)$, its *immediate c[ommand]-string* is $\mathrm{ics}(n) = d_1 \cdots d_i \, n$. The *augmented c[ommand]-string* $\mathrm{acs}(n)$ of n is recursively defined as shown below, where $\uparrow$ is a distinguished symbol:

$$\mathrm{acs}(n) := \begin{cases} \mathrm{ics}(n) & \text{if } n \text{ is the root of } t \\ \mathrm{acs}(m) \uparrow \mathrm{ics}(n) & \text{if } m \text{ is } n\text{'s mother} \end{cases}$$

Example 4. The c-string of the :: $\mathrm{N^+D^-}$ in Fig. 3 is ε :: $\mathrm{T^+wh^+C^-}$ $\uparrow$ does :: $\mathrm{V^+T^-}$ $\uparrow$ like :: $\mathrm{D^+D^+V^-}$ $\uparrow$ a :: $\mathrm{N^+D^-}$ the :: $\mathrm{N^+D^-}$

Licensing conditions are then formalized as constraints on the shape of permissible c-strings. This is comparable to restricting a tree via its path language, except that paths are now replaced by c-strings.

Definition 13 (C-string constraints). A c-string constraint C is some string language L over $\Sigma \cup \{\uparrow\}$. A Σ-tree t is well-formed with respect to C iff $\mathrm{acs}(n) \in L$ for every node n of t.

Which subregular class provides the most appropriate fit for syntactic licensing conditions is still a matter of debate. Graf and Shafiei (2019) propose IO-TSL as a generous upper bound. IO-TSL is a subregular string language recently defined in (Graf and Mayer, 2018) in their analysis of Sanskrit n-retroflxion. Shafiei and Graf (2019), on the other hand, argue that at least island constraints are best captured by IBSP, another class from subregular work on phonology (Graf, 2017, 2018a). Neither class is particularly well-understood at this point. Intuitively, IO-TSL treats local dependencies as primitive and reduces non-local constraints to local ones over enriched representations. IBSP, on the other hand, takes all constraints to be non-local and then uses locality domain to prune down their reach. Either way there is ample evidence that all attested conditions that can be correctly stated over c-strings are at most regular. That's all we need to show that they can be enforced by an STA.

Before we proceed, the reader should take note that c-strings as defined in Graf and Shafiei (2019) do not quite capture the standard notion of c-command over phrase structure trees. First of all, movement is factored out, so that c-strings only capture the c-command relations between the base positions where arguments enter the derivation. Graf and Shafiei (2019) point out several options for adding movement, but they do not explore them in depth. We will provide our own STA account for movement interactions later on (§6.1). Another, less important deviation from standard c-command pertains to the status of heads and specifiers. If one interprets linear precedence in command strings as c-command, then specifiers do not c-command their selecting head even though they c-command the head's object. At the same time, the head c-commands the specifier. While there seem to be no cases in the syntactic literature where this difference to c-command matters, it nonetheless highlights that c-strings only approximate the standard notion of c-command.

This approximation of c-command is very convenient for our purposes, though. The construction of a node's c-string closely mirrors the definition of a node's spine, so it is perhaps unsurprising that every regular c-string constraint can be enforced by an STA. The construction of an STA automaton for this purpose is remarkably straightforward.

First, we simplify c-string constraints by considering only constraints that generate prefix-closed sets of c-strings. The lemma below establishes that this assumption is innocuous for determining tree well-formedness.

Lemma 4. *Let L be some regular language of well-formed c-strings, and let L_p be the largest subset of L such that $u \notin L_p$ entails $uv \notin L_p$ for all $u \in (\Sigma \cup \{\uparrow\})^+$ and $v \in (\Sigma \cup \{\uparrow\})^*$. Then a Σ-tree is well-formed with respect to L iff it is well-formed with respect to L_p.*

Proof. We only consider c-strings that do not start or end with $\uparrow$ as these never occur in any trees to begin with. Since $L_p \subseteq L$ and, by Def. 13, a tree is well-formed with respect to c-string set C iff all its c-strings are members of C, two entailments follow immediately: I) if t is well-formed with respect to L_p, it is well-formed with respect to L, and II) if t is ill-formed with respect to L, it is ill-formed with respect to L_p.

Next, suppose t is well-formed with respect to L. Since it contains no illicit c-string, t could only be ill-formed with respect to L_p if it contains some licit c-string uv such that $u \notin L_p$, wherefore $uv \notin L_p$. But if uv is a c-string of t, then so is every non-empty prefix of uv that does not end in $\uparrow$. By our initial assumption, t is well-formed, and hence every non-empty prefix of uv is a member of L. It then must also be a member of L_p because, by definition, L_p must be largest among the prefix-closed subsets of L. It follows that t is well-formed with respect to L_p, too.

Finally, consider the case where t is ill-formed with respect to L_p. Then there is some c-string u of t such that $u \notin L_p$ but every non-empty prefix of u (that does not end in $\uparrow$) is a member of L_p. In this case it must also hold that $u \notin L$, for otherwise L_p is either not largest or violates prefix closure. Consequently, t is also ill-formed with respect to L. $\square$

Intuitively, Lemma 4 capitalizes on the fact that whenever a tree contains at least one unlicensed node, the status of other nodes no longer matters because the tree is already ill-formed. Hence we may freely assume that the c-string of node n is illicit as soon as a prefix of that c-string is illicit, even in cases where n itself would be licensed.

Now let $D := \langle \Sigma \cup \{\uparrow\}, Q, q_I, F, \delta \rangle$ be the complete, deterministic finite-state string automaton that recognizes L_p as defined in Lemma 4. As D is deterministic, it has a unique initial state q_I. Its transition relation $\delta : (Q \times \Sigma) \times Q$ is a function. We expand δ to strings such that $\delta(q, \sigma_1 \sigma_2 \cdots \sigma_n) := \delta(\cdots \delta(\delta(q, \sigma_1), \sigma_2) \cdots, \sigma_n)$. Since L_p is prefix-closed, it also holds that $Q = F \cup \{s\}$, where $s \notin F$ is some sink state from which no other state can be reached except s itself. Prefix closure also entails that $q_I \in F$, so that empty c-strings are always allowed (since by definition c-strings are never empty, this is innocuous).

We then construct the corresponding STA

$A_D := \langle \Sigma, Q, q_I, \Delta \rangle$. For $n \geq 1$, each sensing rule in Δ is of the form

$$q(\sigma^{(n)}(\vec{\sigma_1}, \vec{\sigma_2}, \ldots, \vec{\sigma_n})) \rightarrow$$
$$\sigma(\delta(q, \sigma \uparrow)(\vec{\sigma_1}),$$
$$\delta(q, \sigma \uparrow \sigma_1)(\vec{\sigma_2}),$$
$$\ldots,$$
$$\delta(q, \sigma \uparrow \sigma_1 \sigma_2 \cdots \sigma_{n-1})(\vec{\sigma_n})))$$

where $q \in F$ and, as previously defined at the beginning of §2.2, the use of $\vec{\sigma_i}$ with some symbol $\sigma_i^{(n)}$ is a shorthand for $\sigma_i(x_{i_1}, \ldots, x_{i_n})$. For leaf nodes, we require $q(\sigma^{(0)}) \rightarrow \sigma^{(0)} \in \Delta$ iff both $q \in F$ and $\delta(q, \sigma) \in F$.

This construction effectively simulates runs of the string automaton D over c-strings. The only complication is that the process of assigning a state to σ_i does not consider σ_i itself. But σ_i does affect the states of its right siblings and all its daughters. And if σ_i is a leaf, it indirectly determines whether the state can be removed so that the tree may be accepted.

Theorem 5. *Let L be a regular language of well-formed c-strings. Then there is some STA A such that $L(A)$ is the set of all Σ-trees that are well-formed with respect to L.*

Proof. Following Lemma 4, we replace L by L_p and consider the complete, deterministic automaton D that generates L_p. We construct A from D in the manner described above. We then give a proof by induction on the depth of Σ-trees.

Pick some Σ-tree t and suppose $t \in \Sigma^{(0)}$. The only c-string of t is t. Then A recognizes t iff $q_I \in F$ and $\delta(q_I, t) \in F$. The former holds by definition, so $t \in L(A)$ iff $t \in L(D)$. This establishes the base case.

Next, consider any arbitrary configuration $q(\sigma^{(n)}(\sigma_1, \ldots, \sigma_n))$. By our induction assumption, $q = \delta(q_I, u)$ with $\mathrm{acs}(\sigma) = u\sigma$. Suppose $\mathrm{acs}(\sigma_i) \notin L_p$ for some $1 \leq i \leq n$. Then $\delta(q_I, \mathrm{acs}(\sigma_i)) = \delta(q, \sigma \uparrow \sigma_1 \cdots \sigma_i) \notin F$. If $i < n$, A will assign some non-final state to σ_{i+1}. If $i = n$ and σ_i is not a leaf, A assigns the non-final state to the leftmost daughter of σ_i. In both cases, t now contains some node with a non-final state. But there is no sensing rule with a non-final state on its left-hand side, so that A correctly rejects t. If $i = n$ and σ_i is a leaf, then $\delta(\delta(q, \sigma \uparrow \sigma_1 \cdots \sigma_{i-1}), \sigma_i)$ is not final and consequently there is no suitable leaf rewrite rule. As

this covers all possible configurations for σ_i, we conclude that A rejects every tree that is ill-formed with respect to L_p.

In the other direction, suppose A rejects t. Then there must be some configuration $\sigma(q_1(\sigma_1), \ldots, q_i(\sigma_i), \ldots, q_n(\sigma_n))$ such that q_i is not a final state. But then $\mathrm{acs}(\sigma_{i-1}) \notin L_p$, as desired. $\qquad\square$

Theorem 5 establishes that the same subregular machinery of STAs can be used for both the structure-building operations Merge and Move and the c-command licensing conditions that deeply permeate syntax. The STA perspective also provides a new answer as to why c-command should play such an important role in syntax. How an STA treats a given node depends solely on the spine of that node. The spine contains the node itself, all its ancestors, and the siblings of all the nodes in the spine. This immediately precludes generalized notion of command like the S-command relation of Barker and Pullum (1990), which in modern terminology would be CP-command: x CP-commands y iff x does not reflexively dominate y (or the other way round) and every CP that properly dominates x properly dominates y. As x and y can be arbitrarily deep within distinct subtrees while S-commanding each other, this is not an STA-recognizable command relation. C-command, on the other hand, stays within the narrow confines of STAs.

Admittedly STAs could also selectively ignore some c-commanders, operate with "inverse c-command" where a complement c-commands into its specifiers, or switch between different notions of c-command based on some *modulo* counting condition. So just as with the SpIC, the power of STAs goes quite a bit beyond what is desirable for c-command. Still, it is striking that c-command is a very natural relation from STA perspective, whereas more global notions of command are correctly ruled out. As long as one is willing to accept dependency structures as a natural representation that arises from head-argument relations, c-command is a natural companion of STA-recognizability.

A lot of work remains to be done, though. As just discussed, STAs allow for some very unnatural command relations. Even more troubling is that any arbitrary regular constraint over c-strings can be enforced by an STA. Seeing how syntactic constraints are very limited in the shape of dependencies they enforce, STAs are overly powerful. Hence STAs can only act as an upper bound.

At the same time, our current STA approach is too limited. In cases where licensing conditions interact with movement, the licensing element might not be part of the c-string of a node. Our STA construction, which is based purely on c-strings, will necessarily fail in these cases. In addition, some cases of licensing involve generalized notions of c-command that go beyond what STAs can handle. We are confident, though, that these issues can be addressed in future work. The next section briefly sketches the solutions we have in mind.

6 Expanding the Core Results

6.1 Interactions of Movement and Licensing

Linguists have identified many cases where movement obfuscates licensing configurations by displacing the licensed element. A simple example would be *[which book about himself]$_i$ does John like t_i*, where the reflexive is not c-commanded by its antecedent *John* in the corresponding phrase structure tree. These cases are entirely unproblematic for STAs since the MG dependency trees fully factor out movement, so *which book about himself* remains in the object position where it is c-commanded by the subject *John*. The problematic cases are much rarer, to such a degree that convincing examples are hard to come by: I) movement of a licensed element bleeding licensing, and II) movement of a licensing element feeding licensing.

The first case covers configurations where a licensed element — or a subtree containing it — moves to a position above the element's licensor and where the subsequent change in c-command relations does make licensing impossible (in contrast to the binding example above, where licensing holds nonetheless). In MGs with the SpIC this can be captured by an STA. The states of the STA would be n-tuples similar to the output of the feat-function we defined for MGs. The first component records c-string states in the usual fashion, whereas each other component i records the most recent head h with an f_i^+-feature, plus the state that was assigned to h. When an f_i-mover m is found, the information in the i-th component is used to compute the state for m as if m were a left daughter of h. In this case, m is also excluded from the state computation of its siblings in the dependency

tree. The details remain to be worked out, but movement of a licensee should be easy enough to handle because the STA can separately keep track of the c-command configurations for each position that is targeted by a mover.

The second option is more complicated. Here some phrase must move into a higher position from where it can c-command the element that must be licensed. In combination with the SpIC, this means that the licensing of a node can be contingent on the nodes contained by its rightmost sibling. An STA can still handle this, but it requires a very different construction from the one described in this paper. As in the bleeding case, a state consists of multiple components, each one of which corresponds to a movement feature. Component i records all the types of licensing conditions that still must be met for nodes that are c-commanded from the most recent head with licensor feature f_i^+. The states also keep track of the c-command relations between the heads hosting the relevant licensor features. When a mover with f_i^- is encountered, the STA checks if the mover can satisfy any of the licensing requirements in component i. If so, those are removed from component i and all other components whose heads are c-commanded by the head for component i. At the end, no state may contain any non-empty components. This strategy reimplements licensing as a mechanism where the STA accrues "licensing debt" while moving through the tree. This debt has to be paid off by movers at a later point. The strategy works because the SpIC allows us to correctly synchronize the licensing debt across the states of all daughter nodes.

While each strategy is relatively simple on its own, integrating them is more difficult. An element may move to a higher position p from where it is only licensed by another element that moves to an even higher position. This requires keeping track of potential licensing debts for p which are then narrowed down to the actual licensing debt once it is known what actually moves to p, and then this debt must be paid off by whatever moves to a position above p. Further complicating the picture, some licensing requirements only need to be satisfied once (e.g. Principle A in the example above), whereas others hold throughout the derivation and are enforced after each movement step. The individual components of the automaton states thus must be synchronized in just the right way, which complicates the construction of the STA even more.

6.2 Subcommand

It has been argued that some cases of long-distance binding involve *subcommand* instead of c-command (see Tang 1989 and Huang and Liu 2001, a.o.). A node x subcommands y iff x c-commands y or x is a specifier of some z that c-commands y. From the perspective of c-strings, x c-commands y iff there is some z in the c-string of y such that $x = z$ or x is the left sibling of a daughter of z. For instance, if Principle A in English allowed for subcommand instead of just c-command, then *John's picture pleases himself* would be well-formed as *John* is a specifier of the subject DP and thus subcommands *himself*.

Subcommand is beyond the reach of STAs because it makes the status of a node n dependent on the daughters of n's left siblings. Similar to the case of movement interactions, there are two possible replies to this. One could point out the rarity of subcommand, and that in the few cases where it arises, it serves as a means to furnish additional antecedents for reflexives beyond those that are already provided via c-command. Hence subcommand might be limited to the syntax-semantics interface and may not directly factor into licensing. Alternatively, one could simply increase the look-ahead of STAs from 1 to 2 so that the daughters of daughters are also taken into account. Subcommand then is just a more demanding case of c-command.

6.3 Extension to MG Derivation Trees

Once one equips STAs with a more powerful look-ahead mechanism to handle subcommand, MG derivation trees once again become a viable alternative to MG dependency trees. All the results in this paper extend from dependency tree to derivation trees if one generalizes STAs to deterministic top-down tree automata with finite look-ahead. That is not surprising because dependency trees can be converted to derivation trees by a linear tree transduction. This shows that derivation trees are the result of separating mothers and daughters in a dependency tree by a finitely bounded amount of material (Merge and Move nodes). Finite look-ahead is a means to reconstruct the relations of the dependency tree that have been obfuscated by this additional material.

It remains to be seen which one of the two representation formats ultimately provides for more insightful characerizations. One issue that derivation trees might shed some light on is the monotonic nature of c-command. With an STA over dependency trees, the daughters could be evaluated in any order to determine the states for a c-string constraint. While our current model proceeds left-to-right, we could have just as well gone right-to-left, inside out, or switched between those options based on the label of the mother. Yet only the first option corresponds to c-command as we know it. In a derivation tree, all information is conveyed via dominance. A specifier is not a sibling of the complement but rather resides in a structurally higher position (cf. the positions of *John* and the CP-complement in Fig. 2). Among all the command variants we just described, the empirically attested one in the form of c-command is the only one that is monotonic with respect to dominance.

6.4 Adjuncts

Another problem of dependency trees is the status of adjuncts. In order to obtain the correct c-command relations, an adjunct of node n would have to be treated as a left sibling of all the specifiers of n. But since adjunction is unbounded, this would mean that a node can have arbitrarily many daughters, whereas STAs are usually defined for trees with a finitely bounded arity. Adjuncts in derivation trees, on the other hand, do not create such issues because each adjunct grows the tree vertically rather than horizontally.

Either way, adjuncts must be subject to the same restriction as specifiers to preserve recognizability by STAs or a suitably generalized variant: even though an adjunct may move on its own, nothing may move out of an adjunct. This is of course a well-established property of adjuncts, known as the Adjunct Island Constraint. Once again, then, our specific subregular perspective derives a well-known limitation of movement.

The Adjunct Island Constraint has some principled exceptions such as *[which car]$_i$ did John drive Mary crazy while trying to fix t_i* (Truswell, 2007). It remains to be seen how these exceptions can be reconciled with our approach.

6.5 Across-the-Board Movement

Another well-known island constraint is the Coordinate Structure Constraint (CSC), which forbids extraction from a conjunct. The CSC itself is easy enough to enforce with an STA. It is the exception to the CSC that is of interest here: extraction from a conjunct is permitted if extraction takes place from all conjuncts in the same coordination. Hence *which beer did Ed buy and Greg drink* is well-formed even though *which beer did Ed buy wine and Greg drink* is illicit. This is known as across-the-board movement (ATB).

Curiously, the ATB-exception is very natural from the STA perspective. STAs fail if one cannot clearly tell from the local configuration which one of several subtrees a move feature should be passed into. The SpIC and the Adjunct Island Constraint address this by excluding specifiers and adjuncts from this equation, leaving the complement as the only subtree that might contain additional movers. ATB-movement constitutes the opposite solution where the feature is instead passed into every subtree. This, too, is a fully deterministic process and thus within the limits of STAs. If even one conjunct did not need to contain a mover, then STAs would be faced with a non-deterministic choice that they cannot handle. While this has to be explored in greater detail based on a proper formalization of ATB-movement (e.g. Torr and Stabler 2016), it is remarkable that the abilities of STAs closely line up with the attested movement configurations.

6.6 Connection to Parsing

The preceding discussion shows that STAs not only furnish a tighter upper bound on syntactic complexity, they also explain core aspects of syntax: the importance of island constraints, the availability of ATB-movement, and the central role of c-command, which merely co-opts mechanisms that are independently needed for movement. But this in turn raises the fundamental question why STA-recognizability should be a relevant concept in syntax. We conjecture that STAs exhibit two properties that are attractive for parsing: top-down recognition, and determinism.

The highly predictive nature of human sentence processing suggests that some top-down strategy is employed, either directly in the form of recursive descent parser, or as a top-down filter of a left-corner parser. Given a choice between bottom-up or top-down recognition, the latter is a much more natural match for such parsing algorithms. Precompiling a top-down filter into the parse schema of an MG parser like the one in Stabler (2011b)

or Stanojević and Stabler (2018) is not trivial, but feasible. Determinism ensures that this precompilation does not explode the parse space, which would slow down the parser. Hence STAs are a good match for current MG models of human sentence processing (Kobele et al., 2013; Gerth, 2015; Graf et al., 2017).

Of course a deterministic top-down automaton would also exhibit these properties, but these automata cannot even handle Merge, let alone Move. STAs are a minimal extension of deterministic top-down automata while also furnishing plenty of power for syntactic dependencies. They require some restrictions on movement, but each one of them also improves parsing performance by exponentially reducing the search space (see the discussion of the SpIC's impact on parsing performance in Stabler 2013).

If one assumes that the grammar is but a high-level description of the parser, the restriction to STAs may be an abstract counterpart to various parser constraints that are meant to improve efficiency. Subregular complexity in syntax may thus be closely connected to parsing.

7 Conclusion

We have shown that all current results on the subregular complexity of syntax are insightfully subsumed by sensing tree automata operating over MG dependency trees. The limits of STA-recognizability line up closely with well-established restrictions on movement and syntactic licensing conditions. A lot of issues remain to be formally worked out, but we are confident that the perspective developed in this paper will greatly expand our understanding of subregular syntax.

Acknowledgments

The work reported in this paper was supported by the National Science Foundation under Grant No. BCS-1845344. We thank the participants of the 2019 workshop on subregular complexity at Stony Brook University for their feedback. We are also grateful to the three anonymous reviewers, whose suggestions allowed us to streamline a lot of the formal definitions. In particular Reviewer 3 went above and beyond the call of duty.

References

Rolf Backofen, James Rogers, and K. Vijay-Shanker. 1995. A first-order axiomatization of the theory of finite trees. *Journal of Logic, Language and Information*, 4:5–39.

Hyunah Baek. 2018. Computational representation of unbounded stress: Tiers with structural features. In *Proceedings of CLS 53*, pages 13–24.

Chris Barker and Geoffrey K. Pullum. 1990. A theory of command relations. *Linguistics and Philosophy*, 13:1–34.

Patrick Blackburn, Claire Gardent, and Wilfried Meyer-Viol. 1993. Talking about trees. In *Proceedings of the Sixth Conference of the European Chapter of the Association for Computational Linguistics*, pages 30–36.

Marisa Ferrara Boston, John T. Hale, and Marco Kuhlmann. 2010. Dependency structures derived from Minimalist grammars. In Christian Ebert, Gerhard Jäger, and Jens Michaelis, editors, *MOL 10/11*, volume 6149 of *Lecture Notes in Computer Science*, pages 1–12. Springer, Berlin.

Noam Chomsky. 1995. *The Minimalist Program*. MIT Press, Cambridge, MA.

H. Comon, M. Dauchet, R. Gilleron, C. Löding, F. Jacquemard, D. Lugiez, S. Tison, and M. TommasiK. 2008. Tree automata: Techniques and applications. Published online: http://www.grappa.univ-lille3.fr/tata. Release from November 18, 2008.

Thomas Cornell and James Rogers. 1998. Model theoretic syntax. In *The Glot International State of the Article Book*, volume 1 of *Studies in Generative Grammar 48*, pages 101–125. Mouton de Gruyter.

Robert Frank and K Vijay-Shanker. 2001. Primitive c-command. *Syntax*, 4(3):164–204.

Sabrina Gerth. 2015. *Memory Limitations in Sentence Comprehension. A Structure-Based Complexity Metric of Processing Difficulty*. Ph.D. thesis, University of Potdsam.

Saul Gorn. 1967. Explicit definitions and linguistic dominoes. In *Systems and Computer Science, Proceedings of the Conference held at University of Western Ontario, 1965*, Toronto. University of Toronto Press.

Thomas Graf. 2012a. Locality and the complexity of Minimalist derivation tree languages. In *Formal Grammar 2010/2011*, volume 7395 of *Lecture Notes in Computer Science*, pages 208–227, Heidelberg. Springer.

Thomas Graf. 2012b. Movement-generalized Minimalist grammars. In *LACL 2012*, volume 7351 of *Lecture Notes in Computer Science*, pages 58–73.

Thomas Graf. 2017. The power of locality domains in phonology. *Phonology*, 34:385–405.

Thomas Graf. 2018a. Locality domains and phonological c-command over strings. In *NELS 48: Proceedings of the Forty-Eighth Annual Meeting of the North East Linguistic Society*, volume 1, pages 257–270, Amherst, MA. GLSA.

Thomas Graf. 2018b. Why movement comes for free once you have adjunction. In *Proceedings of CLS 53*, pages 117–136.

Thomas Graf and Connor Mayer. 2018. Sanskrit n-retroflexion is input-output tier-based strictly local. In *Proceedings of SIGMORPHON 2018*, pages 151–160.

Thomas Graf, James Monette, and Chong Zhang. 2017. Relative clauses as a benchmark for Minimalist parsing. *Journal of Language Modelling*, 5:57–106.

Thomas Graf and Nazila Shafiei. 2019. C-command dependencies as TSL string constraints. In *Proceedings of the Society for Computation in Linguistics (SCiL) 2019*, pages 205–215.

Jeffrey Heinz, Anna Kasprzik, and Timo Kötzing. 2012. Learning in the limit with lattice-structured hypothesis spaces. *Theoretical Computer Science*, 457:111–127.

Jeffrey Heinz, Chetan Rawal, and Herbert G. Tanner. 2011. Tier-based strictly local constraints in phonology. In *Proceedings of the 49th Annual Meeting of the Association for Computational Linguistics*, pages 58–64.

Cheng-Teh James Huang and Cheng-Sheng Luther Liu. 2001. Logophoricity, attitudes and *ziji* at the interface. In Peter Cole, Gabrielle Herman, and Cheng-Tea James Huang, editors, *Long Distance Reflexives*, volume 33 of *Syntax and Semantics*, pages 141–195. Academic Press, New York.

M. A. C. Huybregts. 1984. The weak adequacy of context-free phrase structure grammar. In Ger J. de Haan, Mieke Trommelen, and Wim Zonneveld, editors, *Van Periferie naar Kern*, pages 81–99. Foris, Dordrecht.

Adam Jardine and Kevin McMullin. 2017. Efficient learning of tier-based strictly k-local languages. In *Proceedings of Language and Automata Theory and Applications*, Lecture Notes in Computer Science, pages 64–76, Berlin. Springer.

C. Douglas Johnson. 1972. *Formal Aspects of Phonological Description*. Mouton, The Hague.

Ronald M. Kaplan and Martin Kay. 1994. Regular models of phonological rule systems. *Computational Linguistics*, 20(3):331–378.

Gregory M. Kobele. 2006. *Generating Copies: An Investigation into Structural Identity in Language and Grammar*. Ph.D. thesis, UCLA.

Gregory M. Kobele, Sabrina Gerth, and John T. Hale. 2013. Memory resource allocation in top-down Minimalist parsing. In *Formal Grammar: 17th and 18th International Conferences, FG 2012, Opole, Poland, August 2012, Revised Selected Papers, FG 2013, Düsseldorf, Germany, August 2013*, pages 32–51, Berlin, Heidelberg. Springer.

Gregory M. Kobele and Jens Michaelis. 2011. Disentangling notions of specifier impenetrability. In *The Mathematics of Language*, volume 6878 of *Lecture Notes in Artificial Intelligence*, pages 126–142, Berlin, Heidelberg. Springer.

Gregory M. Kobele, Christian Retoré, and Sylvain Salvati. 2007. An automata-theoretic approach to Minimalism. In *Model Theoretic Syntax at 10*, pages 71–80.

Wim Martens. 2006. *Static Analysis of XML Transformation- and Schema Languages*. Ph.D. thesis, Hasselt University.

Wim Martens, Frank Neven, and Thomas Schwentick. 2008. Deterministic top-down tree automata: Past, present, and future. In *Proceedings of Logic and Automata 2008*, pages 505–530.

Kevin McMullin. 2016. *Tier-Based Locality in Long-Distance Phonotactics: Learnability and Typology*. Ph.D. thesis, University of British Columbia.

Jens Michaelis. 2004. Observations on strict derivational minimalism. *Electronic Notes in Theoretical Computer Science*, 53:192–209.

Jens Michaelis. 2009. An additional observation on strict derivational minimalism. In *FG-MOL 2005*, pages 101–111.

Jens Michaelis and Marcus Kracht. 1997. Semilinearity as a syntactic invariant. In *Logical Aspects of Computational Linguistics*, volume 1328 of *Lecture Notes in Artifical Intelligence*, pages 329–345. Springer.

Uwe Mönnich. 2006. Grammar morphisms. Ms. University of Tübingen.

Frank Morawietz. 2003. *Two-Step Approaches to Natural Language Formalisms*. Walter de Gruyter, Berlin.

James Rogers. 1998. *A Descriptive Approach to Language-Theoretic Complexity*. CSLI, Stanford.

James Rogers. 2003. Syntactic structures as multi-dimensional trees. *Research on Language and Computation*, 1(1):265–305.

James Rogers, Jeffrey Heinz, Gil Bailey, Matt Edlefsen, Molly Vischer, David Wellcome, and Sean Wibel. 2010. On languages piecewise testable in the strict sense. In Christan Ebert, Gerhard Jäger, and Jens Michaelis, editors, *The Mathematics of Language*, volume 6149 of *Lecture Notes in Artificial Intelligence*, pages 255–265. Springer, Heidelberg.

Sylvain Salvati. 2011. Minimalist grammars in the light of logic. In Sylvain Pogodalla, Myriam Quatrini, and Christian Retoré, editors, *Logic and Grammar — Essays Dedicated to Alain Lecomte on the Occasion of His 60th Birthday*, number 6700 in Lecture Notes in Computer Science, pages 81–117. Springer, Berlin.

Nazila Shafiei and Thomas Graf. 2019. The subregular complexity of syntactic islands. Ms., Stony Brook University.

Stuart M. Shieber. 1985. Evidence against the context-freeness of natural language. *Linguistics and Philosophy*, 8(3):333–345.

Edward P. Stabler. 1997. Derivational Minimalism. In Christian Retoré, editor, *Logical Aspects of Computational Linguistics*, volume 1328 of *Lecture Notes in Computer Science*, pages 68–95. Springer, Berlin.

Edward P. Stabler. 2011a. Computational perspectives on Minimalism. In Cedric Boeckx, editor, *Oxford Handbook of Linguistic Minimalism*, pages 617–643. Oxford University Press, Oxford.

Edward P. Stabler. 2011b. Top-down recognizers for MCFGs and MGs. In *Proceedings of the 2nd Workshop on Cognitive Modeling and Computational Linguistics*, pages 39–48.

Edward P. Stabler. 2013. Two models of minimalist, incremental syntactic analysis. *Topics in Cognitive Science*, 5:611–633.

Miloš Stanojević and Edward Stabler. 2018. A sound and complete left-corner parser for Minimalist grammars. In *Proceedings of the 8th Workshop on Cognitive Aspects of Computational Language Learning and Processing*, pages 65–74.

Chih-Chen Jane Tang. 1989. Chinese reflexives. *Natural Language and Linguistic Theory*, 7:93–121.

John Torr and Edward P. Stabler. 2016. Coordination in Minimalist grammars: Excorporation and across the board (head) movement. In *Proceedings of the 12th International Workshop on Tree Adjoining Grammars and Related Formalisms (TAG+12)*, pages 1–17, Düsseldorf, Germany.

Robert Truswell. 2007. Extraction from adjuncts and the structure of events. *Lingua*, 117:1355–1377.

Mai Ha Vu. 2018. Towards a formal description of NPI-licensing patterns. In *Proceedings of the Society for Computation in Linguistics (SCiL) 2018*, volume 1, pages 154–163. Article 17.

Mai Ha Vu, Nazila Shafiei, and Thomas Graf. 2019. Case assignment in TSL syntax: A case study. In *Proceedings of the Society for Computation in Linguistics (SCiL) 2019*, pages 267–276.

Presupposition Projection and Repair Strategies in Trivalent Semantics

Yoad Winter
Utrecht University
The Netherlands
y.winter@uu.nl

Abstract

In binary propositional constructions S_1 *con* S_2, the Strong Kleene connectives explain filtering of S_1's and S_2's presuppositions depending on their logical relations with their non-presuppositional content. However, the presuppositions derived by the Strong Kleene connectives are weak conditional presuppositions, which raise the "proviso problem" in cases where no filtering is motivated. Weak Kleene connectives do not face this problem, but only because their presuppositions are often too strong, and hence do not account for filtering phenomena altogether. While various mechanisms have been proposed to allow filtering without the proviso problem, their relations with the standard trivalent Kleene systems have remained unclear. This paper shows that by sacrificing truth-functionality, we uncover a rich domain of possibilities in trivalent semantics in between the Weak Kleene and Strong Kleene connectives. These systems derive presupposition filtering while avoiding the proviso problem. The Kleene-style operators studied are generalized to arbitrary binary functions, which further clarifies the connection between their different "repair" strategies and presupposition projection.

1 Introduction

Logical theories of natural language semantics and pragmatics treat presupposition (Beaver and Geurts, 2014) as a special sort of inference, which in the following examples we denote '$\rightsquigarrow$':

(1) Sue stopped smoking
    $\rightsquigarrow$ Sue used to smoke

  The king of France is bald
    $\rightsquigarrow$ There is a (unique) king of France

  Dan regretted visiting LA
    $\rightsquigarrow$ Dan visited LA

  It was Zoe who stole my car
    $\rightsquigarrow$ Someone stole my car

What makes presuppositions semantically distinguished from other entailments is their special projection properties: presuppositions are preserved under various operators that make other entailments disappear. For instance, the complex sentences in (2)-(4) below embed the sentence *Sue stopped smoking*, whose non-presuppositional entailment *Sue doesn't smoke* is not entailed by any of these complex sentences. By contrast, the statement *Sue used to smoke* is also a presupposition of sentences (2)-(4). In semantic jargon we say that presuppositions are "projected" from conditionals, negation and epistemic modals.[1]

(2) <u>If</u> *Sue stopped smoking* then Dan is happy now.

(3) <u>It is not the case that</u> *Sue stopped smoking*.

(4) <u>Possibly</u>, *Sue stopped smoking*.

We treat such basic projections in trivalent semantics, where natural language sentences are represented using propositions that denote 1 (true), 0 (false) or $*$ (presupposition failure). We say that a proposition φ is *bivalent* if $[\![\varphi]\!]^M \neq *$ for any model M. Basic entailment, presupposition and bivalent-presupposition relations are defined as follows (Keenan, 1973; Beaver, 1997):

Definition 1.1. *For propositions φ and ψ:*

---

[1] When it comes to complex sentences like (2)-(4), the literature is divided on the status of projected presuppositions like *Sue used to smoke*: semantic theories treat projected presuppositions as logically entailed from one reading of the embedding sentence; pragmatic theories (Gazdar, 1979; Chierchia and McConnel-Ginet, 1990) see them as instances of defeasible reasoning. With the rest of the literature on semantic presupposition, we here assume a logical entailment account, embracing a systematic ambiguity of sentences like (2)-(4). Under one reading the presupposition is *suspended*, e.g. using Bochvar's *assertion operator* (Beaver and Krahmer, 2001) or Heim's strategy of *local accommodation* (Heim, 1983). This suspension is highlighted in contexts *it is not the case that Sue stopped smoking, since she never smoked*, which were the center of the Russell-Strawson debate.

$\varphi \Rightarrow \psi \; \equiv \; \varphi$ **entails** ψ *if*

for every model M: if $[[\varphi]]^M = 1$ *then* $[[\psi]]^M = 1$

$\varphi \rightsquigarrow \psi \; \equiv \; \varphi$ **presupposes** ψ *if*

for every model M: if $[[\varphi]]^M \neq *$ *then* $[[\psi]]^M = 1$

$\psi = MBP(\varphi) \; \equiv \; \psi$ *is the* **maximal bivalent presupposition** *(MBP) of* φ *if*

for every model M: $[[\varphi]]^M \neq *$ *iff* $[[\psi]]^M = 1$

To construct elementary presuppositional propositions from bivalent propositions, we employ Blamey's *transplication* operation on bivalent propositions (Blamey, 1986; Beaver and Krahmer, 2001). For bivalent propositions φ_1 and φ_2, the transplication $(\varphi_1 : \varphi_2)$ is defined by:

$$[[(\varphi_1 : \varphi_2)]]^M = \begin{cases} [[\varphi_2]]^M & [[\varphi_1]]^M = 1 \\ * & [[\varphi_1]]^M = 0 \end{cases}$$

For instance, using the bivalent propositions *US* and *S*, we employ the following treatments of simple natural language sentences:

(5) a. Sue used to smoke   *US*

     b. Sue doesn't smoke $\neg S$

     c. Sue stopped smoking $(US : \neg S)$

In this analysis, (5c) entails but doesn't presuppose (5b), and presupposes (hence entails) (5a), as intuitively required. Furthermore, (5a) is the maximal bivalent presupposition of (5c).

For any bivalent φ_1, φ_2 and ψ, the implication operator in the Weak Kleene system (WK, Table 1) satisfies the following:

(WK$_1$) $MBP((\varphi_1 : \varphi_2) \to \psi) \; = \; \varphi_1$

For instance, in sentence (2), property (WK$_1$) correctly accounts for the projection of the presupposition *Sue used to smoke*. Implication in the Strong Kleene system (SK, Table 2) supports a weaker presupposition:

(SK$_1$) $MBP((\varphi_1 : \varphi_2) \to \psi) \; = \; \varphi_1 \vee \psi$

This property means that SK implication expects sentence (2) to presuppose *Sue used to smoke or Dan is happy*, invoking the intuitively irrelevant disjunct *Dan is happy*. This kind of derivation of irrelevant disjuncts in presuppositions is referred to as the "proviso problem" (Geurts, 1996), and appears with all SK connectives (see e.g. (10) below).

¬	
0	1
1	0
*	*

∧	0	1	*
0	0	0	*
1	0	1	*
*	*	*	*

∨	0	1	*
0	0	1	*
1	1	1	*
*	*	*	*

→	0	1	*
0	1	1	*
1	0	1	*
*	*	*	*

Table 1: Weak Kleene (WK) connectives

¬	
0	1
1	0
*	*

∧	0	1	*
0	0	0	0
1	0	1	*
*	0	*	*

∨	0	1	*
0	0	1	*
1	1	1	1
*	*	1	*

→	0	1	*
0	1	1	1
1	0	1	*
*	*	1	*

Table 2: Strong Kleene (SK) connectives

However, as has been often observed (Peters, 1979; Beaver, 1997), other cases of presupposition projection reveal substantial advantages to SK connectives over WK connectives in terms of their linguistic adequacy. Let us consider for example the following sentence:

(6) If Sue used to smoke, she stopped smoking.

Sentence (6), unlike (2), is not felt to presuppose that Sue used to smoke, and similarly sentence (7) below:

(7) If Sue used to smoke Marlboros, she stopped smoking.

In semantic jargon, we say that sentences (6) and (7) are cases of presupposition *filtering*. In these sentences, the antecedent *Sue used to smoke (Marlboros)* entails the presupposition *Sue used to smoke* of the consequent. As a result, that presupposition gets "filtered out" and is not projected as an entailment of the conditional sentence. Such linguistic facts about filtering are accounted for by SK connectives but not by WK connectives. For WK and SK implication, this is exemplified by the following facts for any bivalent φ, ψ_1 and ψ_2:

(WK$_2$) $MBP(\varphi \to (\psi_1 : \psi_2)) \; = \; \psi_1$

(SK$_2$) $MBP(\varphi \to (\psi_1 : \psi_2)) \; = \; \varphi \to \psi_1$

Thus, for sentences (6) and (7), WK implication counter-intuitively expect the MBP to be *Sue used to smoke*. By contrast, (SK_2) correctly expects the MBPs of these sentences to be tautological, which accounts for presupposition filtering.

Conditional MBPs as in the SK system have been argued to also be intuitively correct in cases that do not involve simple filtering (Karttunen and Peters, 1979; Heim, 1983; Beaver, 2001). For instance, let us consider sentence (8) below:

(8) If Sue used to smoke, she stopped smoking Marlboros.

In this conditional sentence, the antecedent is asymmetrically entailed by the consequent. Fact (WK_2) above means that the WK implication operator expects the MBP of (8) to be *Sue used to smoke Marlboros*. This is incorrect, for such an MBP would entail the antecedent in sentence (8), with would counter-intuitively treat the sentence as equivalent to the non-conditional sentence *Sue stopped smoking Marlboros*. To avoid this problem, Karttunen and Peters (1979) and others proposed treatments where the MBP of sentence (8) is as paraphrased below:

(9) If Sue used to smoke, she used to smoke Marlboros.

When analyzed as a material implication, this conditional statement is also what fact (SK_2) about Strong Kleene implication expects as the MBP of sentence (8).

To summarize, while both the WK implication and the SK implication deal with basic projection problems, they are facing complementary difficulties. The WK implication often "projects too much", failing to filter out presuppositions in the consequent, or at least conditionalize them. In other cases, however, WK implication is advantageous to the conditional presuppositions derived by the SK implication. These SK-based presuppositions are often too weak, and lead to the so-called "proviso problem" for SK implication.

Similar puzzles appear with the other binary propositional connectives in the Kleene truth tables. For instance, the WK and SK conjunction connectives satisfy the following:

$(WK_3) \quad MBP(\varphi \wedge (\psi_1 : \psi_2)) = \psi_1$

$(SK_3) \quad MBP(\varphi \wedge (\psi_1 : \psi_2)) = \varphi \rightarrow \psi_1$

For instance, with *EX*, *S*/*SM* and *US*/*USM* for "Sue exercises", "Sue smokes (Marlboros)" and "Sue used to smoke (Marlboros)", respectively, this leads to the following analyses of the sentences below:

(10) Sue exercises a lot and stopped smoking
    ⇝ Sue used to smoke

    by (WK_3): $MBP(EX \wedge (US : \neg S)) = US$
        ✓ (no proviso problem)

    by (SK_3): $MBP(EX \wedge (US : \neg S)) = EX \rightarrow US$
        ✗ (proviso problem)

(11) Sue used to smoke Marlboros and stopped smoking
    ⇸ Sue used to smoke

    by (WK_3): $MBP(USM \wedge (US : \neg S)) = US$
        ✗ (no filtering)

    by (SK_3): $MBP(USM \wedge (US : \neg S)) = USM \rightarrow US$
        ✓ (tautological presupposition, hence filtering)

(12) Sue used to smoke and stopped smoking Marlboros
    ⇸ Sue used to smoke Marlboros[2]

    by (WK_3): $MBP(US \wedge (USM : \neg SM)) = USM$
        ✗ (no filtering or conditional presupposition)

    by (SK_3): $MBP(US \wedge (USM : \neg SM)) = US \rightarrow USM$
        ✓ (conditional presupposition)

Table 3 summarizes the theoretical puzzle for the WK and SK connectives. To solve this puzzle, we should like a trivalent account of presupposition projection to avoid proviso-like presuppositions while allowing presupposition filtering and conditional presuppositions. This paper proposes such an account, by considering the possibilities that open up once we renounce the truth-functionality of the WK and SK connectives.

	Filtering	Conditional presuppositions	No Proviso
WK	✗	✗	✓
SK	✓	✓	✗

Table 3: Kleene systems and presupposition projection

---

[2]To see that *Sue used to smoke Marlboros* is not simply presupposed by the sentence in (12), we should look at sentences where such conjunctions are embedded, as in: *if Sue used to smoke and stopped smoking Marlboros, then she might be smoking now other brands of cigarettes*. This makes it clear that the statement *Sue used to smoke Marlboros* is not projected as a presupposition of the conditional, hence, reasonably, also not of the conjunction that it contains.

2 Kleene systems of intermediate strength

This section develops two trivalent systems that on the one hand account for presupposition filtering and conditional presuppositions, and on the other hand avoid the unnecessarily weak presuppositions that lead to the "proviso problem" with SK connectives. This account involves employing inferential relations between arguments of binary propositional operators. While such inferences between operands have occasionally been employed under various assumptions (Beaver, 1999; Lassiter, 2012; Mandelkern, 2016), we aim here to employ them within a purely trivalent semantics that allows a better insight the role of Kleene connectives in natural language semantics, in search for a linguistically adequate "intermediate" trivalent semantics in between WK and SK.[3]

2.1 Entailment relations and presupposition filtering

Let us first reconsider the contrast between sentences (10) and (11), which are restated below:

(10')   Sue exercises a lot and stopped smoking.

(11')   Sue used to smoke Marlboros and
        stopped smoking.

As we saw, sentence (10) intuitively presupposes that *Sue stopped smoking*, whereas sentence (11) does not. This kind of difference in filtering is often analyzed in terms of whether the presupposition of the second conjunct is entailed by the first conjunct (Mandelkern, 2016). In example (11) the first conjunct *Mary used to smoke Marlboros* entails the MBP *Sue used to smoke* of the second conjunct. Such an entailment is missing in (10). These facts are used as the source of filtering in (11) and the lack thereof in (10).

Formalizing this filtering principle in trivalent semantics, we get the following restriction on the interpretation of $\varphi \wedge \psi$:

(13) Left-to-right filtering in conjunctions $\varphi \wedge \psi$:

If $\varphi \Rightarrow \mathrm{MBP}(\psi)$, then $\mathrm{MBP}(\varphi) \Rightarrow \mathrm{MBP}(\varphi \wedge \psi)$

In words: if the left-hand conjunct φ in a conjunction $\varphi \wedge \psi$ entails the maximal presupposition of the right-hand conjunct ψ, then that presupposition gets "filtered out", i.e. all presuppositions of $\varphi \wedge \psi$ are inherited from φ. In example (11), the left-hand conjunct (*Sue used to smoke*) is bivalent, hence (13) correctly expects the conjunction to also be bivalent. This accounts for the "filtering" of the presupposition in the right-hand conjunct. At the same time, (13) on its own does not expect filtering in (10), where the entailment $\varphi \Rightarrow \mathrm{MBP}(\psi)$ does not hold.

As illustrated by the WK analysis of sentence (11) (section 1), WK conjunction does not satisfy the condition in (13), hence its failure to account for filtering phenomena in such sentences. By contrast, the following fact about SK conjunction makes it clear that it does satisfy (13):

(14) $[[\mathrm{MBP}(\varphi \wedge_{\mathrm{SK}} \psi)]]^M =$

$$\begin{cases} 1 & ([[\mathrm{MBP}(\varphi)]]^M = 1 \text{ and } [[\mathrm{MBP}(\psi)]]^M = 1) \\ & \text{or } [[\varphi]]^M = 0 \text{ or } [[\psi]]^M = 0 \\ 0 & \text{otherwise} \end{cases}$$

Fact (14) about SK conjunction leads to its desirable filtering property, but it also leads to proviso problems as in (10), for there are cases where the entailment $\varphi \Rightarrow \mathrm{MBP}(\psi)$ does *not* hold, but SK conjunction admits models where $[[\varphi \wedge \psi]]^M$ is bivalent although $[[\psi]]^M = *$ – namely, the models M where $[[\varphi]]^M = 0$.

To address these problems of the WK and SK systems, it is useful to first observe their take on the following question:

(Q) Let op_2 be a *bivalent* binary propositional operator, and let op_3 be the corresponding *trivalent* operator. Which formulas φ, ψ and models M admit a bivalent value for $[[\varphi \, op_3 \, \psi]]^M$ when $[[\varphi]]^M = *$ or $[[\psi]]^M = *$?

The WK system treats the value '*' as "nonsense", and accordingly, its answer on (Q) is "no formulas, and no models".

The SK system treats the value '*' as "unknown", and uses the fact that certain values of an argument of a binary function may determine the result of the function regardless of the value

---

[3]The present modification of Kleene systems is orthogonal to the familiar proposal in (Peters, 1979), where the "intermediate" Kleene system is aimed to treat left-right asymmetries of presupposition projection with binary connectives (Mandelkern et al., 2017). This kind of asymmetry does not concern "proviso problems" that result from presuppositions of the righthand operand, which are the focus of the present paper. Peters's asymmetries are introduced in the current non-truth-functional proposal (see note 6 below), but if needed they can also be removed.

of the other argument. For the standard bivalent connectives, these "decisive values" are 0 for both operands of conjunction, 1 for both operands of disjunction, and 0/1 respectively for the left-hand/righthand operand of implication.[4] The answer of the SK system to (Q) may then be expressed as follows:

(A$_1$) **SK's answer on (Q)**: All formulas, and any model where the value $[[\varphi]]^M$ or $[[\psi]]^M$ determines the result of op_2.

(*Motivation*: extract as much information as possible from known values)

The proviso problem demonstrates that for natural language, the answer in (A$_1$) is too liberal. The problem lies in the fact that the SK answer allows "saving" a formula $\varphi \, op_3 \, \psi$ from having a '$*$' value in some model, with no respect to whether the formula can also be "saved" in the same way in other models. Thus, supposing that the second conjunct in sentence (10) involves a presupposition failure, we see that SK incorrectly "saves" the conjunction from failure if the first conjunct is false. At the same time, SK correctly treats such a conjunction as a failure in models where the first conjunct is true. We consider this "global instability" of the way failures are handled in SK as the source of the proviso problem. Instead of (A$_1$), we propose a "globally stable" variant of SK's answer to (Q). Since this answer minimally strengthened WK, we refer to the system on which it is based as 'WK$^+$'. The "WK$^+$ answer" is informally stated below:

(A$_2$) **WK$^+$ answer on (Q)**: Only formulas where a failure of one operand guarantees that the other operand also fails, or else has a value that determines the result of op_2.

(*Motivation*: extract as much information as possible from known values in formulas that can be globally saved from failure)

This answer, put informally here, summarizes a common linguistic intuition about the contrast between sentences (10) and (11). In sentence (11) it is guaranteed than whenever the second conjunct fails, the first conjunct is *false*. This is the value that determines the result of bivalent conjunction, hence can "save" the formula from failing.[5] There is no such guarantee for sentence (10). Thus, answer (A$_2$) employs the general SK reasoning, but only for "saving", or "repairing", some of the presupposition failures that SK addresses: those failures that can be globally saved from failing the formula (or, using another metaphor: can be globally "repaired").

Following this reasoning, in (15) below we define a conjunction operator that, like SK conjunction, satisfies the condition in (13), but without the general property (14). The operator in (15) "strengthens" WK conjunction to satisfy property (13), hence we refer to it as a *strengthened WK* (WK$^+$) conjunction operator, which is denoted '$\wedge_{WK}$':

(15) **Conjunction in WK$^+$**:

For propositional formulas φ and ψ, with $\mathcal{M}$ a class of models and $M \in \mathcal{M}$ s.t. $[[\varphi]]^M$ and $[[\psi]]^M$ are inductively specified, we define:

$$[[\varphi \wedge_{WK} \psi]]^M =$$

$$\begin{cases} [[\varphi \wedge \psi]]^M & [[\varphi]]^M \neq * \text{ and } [[\psi]]^M \neq * \\ [[\varphi]]^M & [[\psi]]^M = * \text{ and } \forall M' \in \mathcal{M}: \\ & \quad \text{if } [[\psi]]^{M'} = * \text{ then } [[\varphi]]^{M'} \neq 1 \\ * & \text{otherwise} \end{cases}$$

The first clause in definition (15) standardly retains bivalent conjunction. The second clause makes sure to respect the condition (13). [6] An advantage of the $\wedge_{WK}$ operator over SK conjunction is the avoidance of proviso problems as in (10): falsity of φ entails falsity of $\varphi \wedge_{WK} \psi$ only if the condition in the second clause of definition (15) holds, which is not the case in (10). Formally, for any bivalent propositions φ, ψ_1 and ψ_2, we observe the following fact on the WK, WK$^+$ and SK conjunction operators:

---

[4]Some bivalent connectives, specifically exclusive disjunction, do not have decisive values. For such connectives, the "WK and SK answers" above give identical results, as they do for formulas like $(\varphi \vee \psi) \wedge \neg(\varphi \wedge \psi)$.

[5]Cases where both conjuncts fail are discussed below.

[6]Note that this clause is asymmetric: it makes WK$^+$ conjunction respect the condition in (13), but not the symmetric condition (if $\psi \Rightarrow \mathrm{MBP}(\varphi)$, then $\mathrm{MBP}(\psi) \Rightarrow \mathrm{MBP}(\varphi \wedge \psi)$). Modifying (15) into a symmetric version is straightforward, but it is questionable if such left-right symmetry (similar to that of the SK connectives) would be empirically motivated, as it would expect right-to-left filtering, which is empirically dubious (Peters, 1979; Mandelkern et al., 2017). Thus, to simplify the presentation in this paper, we here only define operators that derive asymmetric filtering and conditional presuppositions.

(16) Assuming that $\varphi \Rightarrow \psi_1$, we have:

WK✗ $\text{MBP}(\varphi \wedge_{\text{wk}} (\psi_1 : \psi_2)) = \psi_1$

WK$^+$✓ $\text{MBP}(\varphi \wedge_{\text{wk}} (\psi_1 : \psi_2)) = \top$

(17) Assuming that $\varphi \not\Rightarrow \psi_1$, we have:

SK✗ $\text{MBP}(\varphi \wedge_{\text{sk}} (\psi_1 : \psi_2)) = (\neg\varphi) \vee \psi_1$

WK$^+$✓ $\text{MBP}(\varphi \wedge_{\text{wk}} (\psi_1 : \psi_2)) = \psi_1$

The $\top$ symbol standardly refers to the univalent proposition denoting 1 in all models. In (16), the '✓'/'✗' symbols mark the correct/incorrect treatment of filtering in sentences like (11). In (17) they mark the avoidance/retainment of the proviso problem in sentences like (10).

A disadvantage of WK$^+$ conjunction over SK conjunction is that the second clause in definition (15) makes WK$^+$ conjunction non-truth-functional, as it relies on logical relations within the whole class of models $\mathcal{M}$.[7]

The $\wedge_{\text{wk}}$ operator follows the general "repair" strategy of SK conjunction. When the second clause in (15) is met, the assignment of the interpretation of φ to $[\![\varphi \wedge_{\text{wk}} \psi]\!]^M$ is motivated by the wish to preserve the following classical property of bivalent conjunction:

$$[\![\varphi]\!]^M = 0 \;\Rightarrow\; [\![\varphi \wedge \psi]\!]^M = 0.$$

Similarly, the following classical property of material implication motivates the treatment of filtering with conditionals as in (7):

$$[\![\varphi]\!]^M = 0 \;\Rightarrow\; [\![\varphi \rightarrow \psi]\!]^M = 1.$$

With disjunction the motivation is to preserve the following property:

$$[\![\varphi]\!]^M = 1 \;\Rightarrow\; [\![\varphi \vee \psi]\!]^M = 1.$$

This motivation is geared by filtering as in the following disjunctive example, which does not presuppose that *Sue used to smoke*:

(18) Either Sue never smoked Marlboros, or she stopped smoking.

These considerations about filtering with conditionals and disjunction lead to the following def-

initions of the respective WK$^+$ operators, in analogy to (15) above:[8]

(19)   Implication and disjunction in WK$^+$:

$$[\![\varphi \mathbin{\dot\rightarrow}_{\text{wk}} \psi]\!]^M =$$

$$\begin{cases} [\![\varphi \rightarrow \psi]\!]^M & [\![\varphi]\!]^M \neq * \text{ and } [\![\psi]\!]^M \neq * \\ [\![\neg\varphi]\!]^M & [\![\psi]\!]^M = * \text{ and } \forall M' \in \mathcal{M}: \\ & \quad \text{if } [\![\psi]\!]^{M'} = * \text{ then } [\![\varphi]\!]^{M'} \neq 1 \\ * & \text{otherwise} \end{cases}$$

$$[\![\varphi \vee_{\text{wk}} \psi]\!]^M =$$

$$\begin{cases} [\![\varphi \vee \psi]\!]^M & [\![\varphi]\!]^M \neq * \text{ and } [\![\psi]\!]^M \neq * \\ [\![\varphi]\!]^M & [\![\psi]\!]^M = * \text{ and } \forall M' \in \mathcal{M}: \\ & \quad \text{if } [\![\psi]\!]^{M'} = * \text{ then } [\![\varphi]\!]^{M'} \neq 0 \\ * & \text{otherwise} \end{cases}$$

Based on the "WK$^+$ answer" in (A$_2$) above, the reasoning behind all these definitions is general, as is further explored in section 3.

The WK$^+$ conjunction operator, as well as WK$^+$ disjunction and implication, also deals with cases where the first conjunct is not bivalent, as in the following example:

(20) [Sue stopped drinking but continued to smoke two Marlboro packs a day], and [now she has finally also stopped smoking].

Intuitively, sentence (20) presupposes that Sue used to drink. This requirement holds independently of Sue's smoking habits. Definition (15) makes sure that the presupposition *Sue used to drink* of the first conjunct in (20) gets projected, despite the filtering of the presupposition *Sue used to smoke* of the second conjunct. In general: for any proposition φ' that is presupposed by φ, we have φ' presupposed by $\varphi \wedge_{\text{wk}} \psi$ as well.[9]

---

[7]This drawback of WK$^+$ conjunction is shared with other previous "globalist" accounts of the proviso problem (Lassiter, 2012; Mandelkern, 2016). If desirable, it might be removed by couching definition (15) within a possible world semantics, replacing quantification over models by quantification over indices in a given model.

[8]Since the 0 value in the left argument similarly determines the result of both conjunction and material implication, the general principle underlying left-to-right filtering with implication is the same as for conjunction in (13). By contrast, with disjunction, the 1 value determines the result, hence the general principle analogous (13) is: if $\neg\varphi \Rightarrow \text{MBP}(\psi)$, then $\text{MBP}(\varphi) \Rightarrow \text{MBP}(\varphi \vee \psi)$.

[9]*Proof:* Let M be a model where $[\![\varphi \wedge_{\text{wk}} \psi]\!]^M \neq *$. Thus, either clause 1 or 2 of definition (15) is satisfied. Clause 1 trivially entails $[\![\varphi]\!]^M \neq *$. Clause 2 entails $[\![\varphi \wedge_{\text{wk}} \psi]\!]^M = [\![\varphi]\!]^M$, from which we also conclude $[\![\varphi]\!]^M \neq *$. If $\varphi \rightsquigarrow \varphi'$, then in every model M s.t. $[\![\varphi \wedge_{\text{wk}} \psi]\!]^M \neq *$, we have $[\![\varphi']\!]^M = 1$. We conclude that $(\varphi \wedge_{\text{wk}} \psi) \rightsquigarrow \varphi'$.

Definitions (15) and (19) quantify over models in a way that accounts for filtering phenomena as in the following example (Beaver, 1999) :

(21) If Jane takes a bath, Bill will be annoyed that there is no more hot water.

As Beaver notes, while the relation between taking a bath and lack of hot water is by no means logical, in normal conversations the presupposition *there is no more hot water* of the consequent in (21) gets filtered out. In general, this filtering is on a par with the filtering phenomena discussed above, where the relations between conjuncts are logical. However, there is one empirical caveat: an entailment $\varphi \Rightarrow \text{MBP}(\psi)$ in (13) which is not logical but restricted to a designated class of models can be explicitly denied in conversation. For instance, when a given context explicitly denies the relation between taking a bath and lack of hot water, filtering in (21) disappears:

(22) The hot water supply in Bill's place uses gas heating, so that no single person could possibly take a bath that would stop the hot water supply. At present there's some problem with Bill's heating system. Not knowing that, Bill suggests Jane, who is staying at his place, to take a bath whenever she pleases. *If Jane takes a bath, Bill will be annoyed (to hear from her) that there is no more hot water.*

Unlike the use of sentence (21) in an out-of-the-blue context, in the context of (22) sentence (21) does presuppose that *there in no more hot water.* Thus, due to the explicit denial in (22) of any causal relation between Jane's bath and the lack of hot water, filtering does not take place. Using a given class of models $\mathcal{M}$ in definitions (15) and (19), rather than all possible models, allows the filtering mechanism to take into account implicit epistemic assumptions, without getting into the separate question of how these assumptions should be modeled. A similar point is made in (Mandelkern, 2016) in relation to a framework of context-change potentials.

2.2 Conditional presuppositions

The WK$^+$ operators defined above do not expect conditional presuppositions, which were exemplified in sentence (8), restated below:

(8') If Sue used to smoke, she stopped smoking Marlboros.

In this case the MBP of the consequent *Sue used to smoke Marlboros* asymmetrically entails the antecedent. The second clause in the definition of the WK$^+$ implication operator in (19) does not hold in such cases. Accordingly, this operator incorrectly expects the presupposition *Sue used to smoke Marlboros* to be projected in (8). Formally, for any bivalent propositions φ, ψ_1 and ψ_2, we observe the following fact on the SK and WK$^+$ implication operators:

(23) Assuming that $\psi_1 \Rightarrow \varphi$ and $\varphi \nRightarrow \psi_1$, we have:

$$\text{SK}\checkmark \quad \text{MBP}(\varphi \rightarrow_{\text{SK}} (\psi_1 : \psi_2)) = (\neg\varphi) \vee \psi_1$$

$$\text{WK}^+\textbf{✗} \quad \text{MBP}(\varphi \dot\rightarrow_{\text{WK}} (\psi_1 : \psi_2)) = \psi_1$$

The '$\checkmark$'/'✗' symbols mark here the correct/incorrect modelling of conditional presuppositions in sentences like (8).

Treating this kind of problem has led previous work to assume that the MBP of sentences like (8) should be expressed by the following disjunction:

(24) Either Sue never smoked or she used to smoke Marlboros.

Within a trivalent system, this treatment of (8) is generalized using the following condition:

(25) Left-to-right conditional presuppositions in implications $\varphi \rightarrow \psi$:

If $\text{MBP}(\psi) \Rightarrow \varphi$, then:

$$\neg\varphi \vee (\text{MBP}(\varphi) \wedge \text{MBP}(\psi)) \Rightarrow \text{MBP}(\varphi \rightarrow \psi).$$

In words: when the MBP of the consequent ψ in $\varphi \rightarrow \psi$ entails the antecedent φ, the negation of φ satisfies the MBP of $\varphi \rightarrow \psi$, as a possible alternative to the straightforward WK-based presupposition $\text{MBP}(\varphi) \wedge \text{MBP}(\psi)$. Principle (25) correctly makes the disjunction in (24) entail the MBP of sentence (8), as expected by the Strong Kleene system. Indeed, SK implication satisfies (25). However, as in relation to presupposition filtering, this treatment of conditional presuppositions comes at the cost of leading to the proviso problem.

A simple trivalent extension of WK$^+$ derives some of the most typical conditional presuppositions that were addressed in the literature.[10] We

---

[10]There are empirical questions on whether conditional presuppositions are needed at all (Mandelkern, 2016). On the other hand, there are also empirical questions on whether principle (25) can cover all conditional presuppositions (Schlenker, 2011). For space and time limitations I ignore

refer to this extension as *weakened SK* (SK$^-$), and base its behavior on the following answer to question (Q) above regarding the formulas that allow a repair of a presupposition failure:

(A$_3$) **SK$^-$ answer on (Q)**: Only formulas as in WK$^+$ (cf. (A$_2$)) as well as formulas where if one operand has a value that determines the result of op_2, the other operand fails.

> (*Motivation*: as in (A$_2$), plus the additional motivation to extract information from a single known value only when this is globally required in order to save a formula from a failure)

Minimal strengthening of the '$\dot{\rightarrow}_{\text{WK}}$' operator using this principle leads to the following operator, which we denote '$\dot{\rightarrow}_{\text{SK}}$':

(26) Implication in SK$^-$:

> For propositional formulas φ and ψ, with $\mathcal{M}$ a class of models and $M \in \mathcal{M}$ s.t. $[[\varphi]]^M$ and $[[\psi]]^M$ are inductively specified, we define:

$$[[\varphi \dot{\rightarrow}_{\text{SK}} \psi]]^M =$$

$$\begin{cases} [[\varphi \rightarrow \psi]]^M & [[\varphi]]^M \neq * \text{ and } [[\psi]]^M \neq * \\ [[\neg\varphi]]^M & ([[\psi]]^M = * \text{ and } \forall M' \in \mathcal{M}: \\ & \quad \text{if } [[\psi]]^{M'} = * \text{ then } [[\varphi]]^{M'} \neq 1) \\ & \text{or} \\ & ([[\varphi]]^{M'} \neq 1 \text{ and } \forall M' \in \mathcal{M}: \\ & \quad \text{if } [[\varphi]]^{M'} \neq 1 \text{ then } [[\psi]]^{M'} = *) \\ * & \text{otherwise} \end{cases}$$

This definition of SK$^-$ implication agrees with SK implication on conditional presuppositions for sentences like (8), but, like WK$^+$ and unlike SK, does not generate proviso problems. Formally, for any bivalent propositions φ, ψ_1 and ψ_2, we have:

(27) a. Filtering – if $\varphi \Rightarrow \psi_1$:
> SK✓: MBP($\varphi \rightarrow_{\text{SK}} (\psi_1 : \psi_2)$) = ⊤
> SK$^-$✓: MBP($\varphi \dot{\rightarrow}_{\text{SK}} (\psi_1 : \psi_2)$) = ⊤

---

these questions here. I believe that further linguistic work is needed in order to determine if conditional presuppositions, or certain types thereof, should be semantically derived. The SK$^-$ system is only presented here as one natural extension of WK$^+$, with no claims for empirical comprehensiveness.

b. Cond.Pres. – if $\psi_1 \Rightarrow \varphi$:
> SK✓: MBP($\varphi \rightarrow_{\text{SK}} (\psi_1 : \psi_2)$) = $(\neg\varphi) \vee \psi_1$
> SK$^-$✓: MBP($\varphi \dot{\rightarrow}_{\text{SK}} (\psi_1 : \psi_2)$) = $(\neg\varphi) \vee \psi_1$

c. No Proviso – if $\varphi \not\Rightarrow \psi_1$ and $\psi_1 \not\Rightarrow \varphi$:
> SK✗: MBP($\varphi \rightarrow_{\text{SK}} (\psi_1 : \psi_2)$) = $(\neg\varphi) \vee \psi_1$
> SK$^-$✓: MBP($\varphi \dot{\rightarrow}_{\text{SK}} (\psi_1 : \psi_2)$) = ψ_1

This establishes that in cases like (8), SK$^-$ implication shows the desirable properties of SK implication, without the undesirable proviso problem.

The way in which definition (26) quantifies over models accounts for conditional presuppositions that are not triggered by logical entailment, but only due to contextually salient inferential relations, similarly to filtering in sentence (21). For instance, according to (Schlenker, 2011), sentence (28) below has the presupposition in (29):

(28) If John visits his parents for Christmas, his sister too will give them hard time.

(29) If John visits his parents for Christmas, someone (namely John) will give them hard time.

This presupposition is treated here by assuming a contextual entailment from *John visits his parents for Christmas* to *someone will give John's parents a hard time*, which is of course far from being a logical entailment.

The reasoning behind the definition of SK$^-$ implication is also used in the following definitions of conjunction and disjunction:

(30) Conjunction and disjunction in SK$^-$:

$$[[\varphi \wedge_{\text{SK}} \psi]]^M =$$

$$\begin{cases} [[\varphi \wedge \psi]]^M & [[\varphi]]^M \neq * \text{ and } [[\psi]]^M \neq * \\ [[\varphi]]^M & ([[\psi]]^M = * \text{ and } \forall M' \in \mathcal{M}: \\ & \quad \text{if } [[\psi]]^{M'} = * \text{ then } [[\varphi]]^{M'} \neq 1) \\ & \text{or} \\ & ([[\varphi]]^{M'} \neq 1 \text{ and } \forall M' \in \mathcal{M}: \\ & \quad \text{if } [[\varphi]]^{M'} \neq 1 \text{ then } [[\psi]]^{M'} = *) \\ * & \text{otherwise} \end{cases}$$

$$[[\varphi \vee_{\mathrm{SK}} \psi]]^M \; =$$

$$\begin{cases} [[\varphi \vee \psi]]^M & [[\varphi]]^M \neq * \text{ and } [[\psi]]^M \neq * \\ [[\varphi]]^M & ([[\psi]]^M = * \text{ and } \forall M' \in \mathcal{M}: \\ & \quad \text{if } [[\psi]]^{M'} = * \text{ then } [[\varphi]]^{M'} \neq 0) \\ & \text{or} \\ & ([[\varphi]]^{M'} \neq 0 \text{ and } \forall M' \in \mathcal{M}: \\ & \quad \text{if } [[\varphi]]^{M'} \neq 0 \text{ then } [[\psi]]^M = *) \\ * & \text{otherwise} \end{cases}$$

Similarly to SK$^-$ implication, these conjunction and disjunction operators admit conditional presuppositions while avoiding the proviso problem. Thus, when φ, ψ_1 and ψ_2 are bivalent, we get:

If $\psi_1 \Rightarrow \varphi$: $\mathrm{MBP}(\varphi \wedge_{\mathrm{SK}} (\psi_1 : \psi_2)) = (\neg\varphi) \vee \psi_1$.

If $\psi_1 \Rightarrow \neg\varphi$: $\mathrm{MBP}(\varphi \vee_{\mathrm{SK}} (\psi_1 : \psi_2)) = \varphi \vee \psi_1$.

2.3 Summary

We have defined two sets of binary operators, referred to as 'WK$^+$' and 'SK$^-$', which satisfy the following, for any operator *op* and trivalent propositions φ and ψ:

$$\mathrm{MBP}(\varphi \, op_{\mathrm{WK}} \, \psi) \;\Rightarrow\; \mathrm{MBP}(\varphi \, op_{\mathrm{WK}^+} \, \psi)$$

$$\Rightarrow\; \mathrm{MBP}(\varphi \, op_{\mathrm{SK}^-} \, \psi) \;\Rightarrow\; \mathrm{MBP}(\varphi \, op_{\mathrm{SK}} \, \psi)$$

Further, we have shown (wit. (27c), (23), (16)):

$$\mathrm{MBP}(\varphi \, op_{\mathrm{SK}} \, \psi) \;\not\Rightarrow\; \mathrm{MBP}(\varphi \, op_{\mathrm{SK}^-} \, \psi)$$

$$\not\Rightarrow\; \mathrm{MBP}(\varphi \, op_{\mathrm{WK}^+} \, \psi) \;\not\Rightarrow\; \mathrm{MBP}(\varphi \, op_{\mathrm{WK}} \, \psi)$$

This describes a hierarchy where SK/WK operators derive the weakest/strongest presuppositions, respectively. Equivalently, and more in line with common nomenclature, Strong Kleene operators have the strongest "failure conditions" (the negation of their MBPs) whereas the failure conditions of Weak Kleene operators are the weakest. In terms of this "strength", the WK$^+$ and SK$^-$ operators are properly in between the two classical Kleene connectives.

3 Repair and value determination with general binary functions

The key to the proposal in section 2 is in the general specification of "repair" conditions for failures in propositional arguments. These are principles that specify the situations under which a presupposition failure in one of a binary function's arguments may still allow the function to return a value. Following (George, 2008, 2014), we aim to make the reasoning behind our proposal more ex-plicit by generalizing it to arbitrary functions. Unlike George's work, we do not necessarily seek to generalize the Strong Kleene connectives, which lead to the proviso problem, but rather to avoid this problem using intermediate levels of presupposition projection as in the WK$^+$ and the SK$^-$ operators. This section generalizes these operators to arbitrary binary functions.

Given a set $X \subset E$ and an element $*$ in $E - X$, we denote $X^* = X \cup \{*\}$. Following (de Groote and Lebedeva, 2010), we view presupposition failure ($*$) as an "exception", which should be optimally "handled" or "repaired". A *repair strategy* α is a method for defining $f^\alpha : (A^* \times B^*) \to C^*$ for any given binary function $f : (A \times B) \to C$.

The Weak Kleene strategy is "no repair". Thus, $f^{\mathrm{WK}} : (A^* \times B^*) \to C^*$ is defined by:

$$(31) \quad f^{\mathrm{WK}}(x,y) \; = \begin{cases} f(x,y) & x \in A \text{ and } y \in B \\ * & x = * \text{ or } y = * \end{cases}$$

By contrast, Strong Kleene is based on a "maximal repair" strategy. A function $f^{\mathrm{SK}} : (A^* \times B^*) \to C^*$ is capable of "repairing" failures of its left/right whenever the result of f can be determined by the value of the right/left argument alone, respectively. This notion of *left/right* (L/R) *determination* is defined as follows:

Definition 3.1. *For any function* $f : (A \times B) \to C$ *and value* $c \in C$, *we say that:*

A value $a \in A$ **L-determines** f *as* c *if* $\forall y \in B . f(a,y) = c$

A value $b \in B$ **R-determines** f *as* c *if* $\forall x \in A . f(x,b) = c$

Using the notion of L-determination, we define the *L-determination function* $LD_f : A^* \to C^*$ of a function $f : (A \times B) \to C$ as follows:

$$(32) \quad LD_f(x) \; = \begin{cases} c & x \in A \text{ and } x \text{ L-determines } f \text{ as } c \\ * & \text{otherwise} \end{cases}$$

Symmetrically, the *R-determination function* $RD_f : B^* \to C^*$ of f is defined by:

$$(33) \quad RD_f(y) \; = \begin{cases} c & y \in B \text{ and } y \text{ R-determines } f \text{ as } c \\ * & \text{otherwise} \end{cases}$$

Bivalent conjunction, disjunction and implication satisfy:

$$LD_\wedge(0) = RD_\wedge(0) = 0 \qquad LD_\wedge(1) = RD_\wedge(1) = *$$
$$LD_\vee(1) = RD_\vee(1) = 1 \qquad LD_\vee(0) = RD_\vee(0) = *$$
$$LD_\to(0) = RD_\to(1) = 1 \qquad LD_\to(1) = RD_\to(0) = *$$

Using the LD and RD functions, we define a Strong Kleene function $f^{\text{SK}} : (A^* \times B^*) \to C^*$ for any binary function $f : (A \times B) \to C$:

(34) $\quad f^{\text{SK}}(x, y) =$
$$\begin{cases} f(x, y) & x \in A \text{ and } y \in B \\ c & c \in C \text{ and } (LD_f(x) = c \\ & \text{or } RD_f(y) = c) \\ * & \text{otherwise} \end{cases}$$

It will be observed that the standard WK/SK connectives (tables 1 and 2) apply the respective repair strategies (31)/(34) to the bivalent connectives. Like their propositional instantiations in the Kleene tables, the more general strategies (31) and (34) are "local" in that for given values x and y, they completely determine the value $f^\alpha(x, y)$ based on f, x and y. By contrast, the WK$^+$ and SK$^-$ operators of section 2 rely on entailments between propositional formulas, hence they are not local in this sense (as mentioned above, the WK$^+$ and SK$^-$ operators are not truth-functional).

In order to compare the WK and SK strategies (31) and (34) to global generalizations of the WK$^+$ and SK$^-$ operators, we first define global versions of the former. Let M be a model over expressions within a type system for n-place functions and products (e.g. van Benthem (1991)). For any type τ, we standardly denote D_τ^M for the domain of values of type τ in M, allowing partial function values. For any such type and model, we assume that the exceptional value $*$ is not in D_τ^M, and denote $D_\tau^{M^*} = (D_\tau^M)^*$. For an expression exp of type τ, we need to specify an element of $D_\tau^{M^*}$ as the denotation of exp. This element is denoted $[\![exp]\!]^{M^*}$. Globalizing the WK and SK repair strategies above, we assume that F is a binary function expression of type $(\text{a} \bullet \text{b})\text{c}$, and exp_1 and exp_2 are expressions of type a and b, respectively. We assume by induction that for every model $M \in \mathcal{M}$: $[\![F]\!]^M \in D_{(\text{a}\bullet\text{b})\text{c}}^M$, $[\![exp_1]\!]^{M^*} \in D_\text{a}^{M^*}$ and $[\![exp_2]\!]^{M^*} \in D_\text{b}^{M^*}$.

Definition 3.2. *The* **global WK and SK strate-** gies *for* $[\![F(exp_1, exp_2)]\!]^{M^*}$ *are defined by:*

$$[\![F^{\text{WK}}(exp_1, exp_2)]\!]^{M^*} =$$
$$([\![F]\!]^M)^{\text{WK}}([\![exp_1]\!]^{M^*}, [\![exp_2]\!]^{M^*})$$

$$[\![F^{\text{SK}}(exp_1, exp_2)]\!]^{M^*} =$$
$$([\![F]\!]^M)^{\text{SK}}([\![exp_1]\!]^{M^*}, [\![exp_2]\!]^{M^*})$$

For example, let 'mult' denote a the standard binary multiplication operator over real number expressions, and let '/' denote the standard partial division operator over real number expressions. Let $[\![1/r]\!]^{M^*}$ be inductively specified as $*$ in models M^* where $[\![r]\!]^{M^*} = 0$. Considering the Kleene-repaired expression $\mathbf{mult}^\alpha(r, 1/r)$, we observe that any model M s.t. $[\![r]\!]^{M^*} = 0$ satisfies $[\![\mathbf{mult}^{\text{WK}}(r, 1/r)]\!]^{M^*} = *$ whereas $[\![\mathbf{mult}^{\text{SK}}(r, 1/r)]\!]^{M^*} = 0$.

Generalizing the and WK$^+$ and SK$^-$ operators of section 2 involves considering the global strategies they employ. In such global strategies, we need to classify which expressions $F(exp_1, exp_2)$ are "repairable" in cases of failure of exp_2. This classification depends on the denotations of F, exp_1 and exp_2 in different models. We start out with generalizing the WK$^+$ operators. Definition 3.3 below specifies the conditions under which the WK$^+$ strategy is allowed to "repair" failures of exp_2. As in definition 3.2, we are given a binary function expression F of type $(\text{a} \bullet \text{b})\text{c}$, and expressions exp_1 and exp_2 of type a and b, respectively, s.t. for every model $M \in \mathcal{M}$: $[\![F]\!]^M \in D_{(\text{a}\bullet\text{b})\text{c}}^M$, $[\![exp_1]\!]^{M^*} \in D_\text{a}^{M^*}$ and $[\![exp_2]\!]^{M^*} \in D_\text{b}^{M^*}$.

Definition 3.3. *Given a class of models $\mathcal{M}$, let $c \in \bigcap_{M \in \mathcal{M}} D_\text{c}^M$, and suppose that every model $M \in \mathcal{M}$ where $[\![exp_2]\!]^{M^*} = *$ and $[\![exp_1]\!]^{M^*} \neq *$ satisfies: $LD_{[\![F]\!]^M}([\![exp_1]\!]^{M^*}) = c$. Then we say that the expression $F(exp_1, exp_2)$ is* **R-repairable as** c.

In words: an expression $F(exp_1, exp_2)$ is R-repairable when there is a value c of type c shared by all models, and for all models, c is L-determined by the value of exp_1 for F whenever exp_2 fails and exp_1 does not.[11]

Example 1: Let us again consider the expression $\mathbf{mult}(r, 1/r)$. The expression $1/r$ denotes $*$ precisely in those models where the value of r is 0, which L-determines the result of $\mathbf{mult}$ as 0. Thus, $\mathbf{mult}(r, 1/r)$ is R-repairable with respect to stan-

---

[11]For the sake of presentation, this condition is stronger than necessary: we might as well require the value c to only be shared by models where exp_2 fails and exp_1 does not.

dard models of the real numbers.

Example 2: Let us consider the expression $F(r, \sqrt{r})$ over the real numbers, where $F(x, y)$ is defined by 0 for $x = -3$ and by $x + y$ otherwise. In models where $r = -3$, this is the value of the left-hand argument of the expression F, which L-determines the result of the function that F denotes. Accordingly, for $r = -3$ the value of $F^{\text{SK}}(r, \sqrt{r})$ is 0, which repairs the failure of $\sqrt{r}$. However, since the expression $\sqrt{r}$ fails for all negative values of r other than -3, and these values do not L-determine the value of F, the expression $F(r, \sqrt{r})$ is not generally R-repairable.

Example 2 highlights a general difference between the SK repair strategy and the WK^+ repair strategy employed in section 2. The SK connectives deal with failures of the right-hand value in all models where the value of the left-hand value L-determines the result. By contrast, the WK^+ operators only deal with failures of the right-hand argument as long as any such failure entails that the value of the left-hand argument L-determines the result. Thus, our propositional WK^+ operators only deal with failures of ψ formulas in *R-repairable* formulas of the form $\varphi\ op\ \psi$. To generalize this WK^+ strategy, we again let F be a binary function expression of type $(\mathsf{a} \bullet \mathsf{b})\mathsf{c}$, and let exp_1 and exp_2 be expressions of type a and b, respectively. We assume by induction that for every model $M \in \mathcal{M}$: $[\![F]\!]^M \in D^M_{(\mathsf{a}\bullet\mathsf{b})\mathsf{c}}$, $[\![exp_1]\!]^{M^*} \in D^{M^*}_{\mathsf{a}}$ and $[\![exp_2]\!]^{M^*} \in D^{M^*}_{\mathsf{b}}$.

Definition 3.4. *The (global)* $\textbf{WK}^+$ *strategy for* $[\![F(exp_1, exp_2)]\!]^{M^*}$ *is defined by:*

$$[\![F^{\text{WK}^+}(exp_1, exp_2)]\!]^{M^*} =$$

$$\begin{cases} [\![F(exp_1, exp_2)]\!]^M \\ \quad [\![exp_1]\!]^{M^*} \neq * \ and\ [\![exp_2]\!]^{M^*} \neq * \\ \\ c \quad LD_{[\![F]\!]^M}([\![exp_1]\!]^{M^*}) = c \in \bigcap_{M \in \mathcal{M}} D^M_{\mathsf{c}} \\ \quad and\ F(exp_1, exp_2)\ is\ R\text{-}repairable\ as\ c \\ \\ * \quad otherwise \end{cases}$$

The propositional WK^+ operators of section 2 are instances of definition 3.4, which is also applicable to binary functions more generally. For instance, based on the facts in examples 1 and 2 above, we conclude that $\mathbf{mult}^{\text{WK}^+}(r, 1/r) = 0$ when $r = 0$, but $F^{\text{WK}^+}(r, \sqrt{r}) = *$ when $r = -3$, in contrast with $F^{\text{SK}}(r, \sqrt{r}) = 0$. This is by virtue of the R-reparability of $\mathbf{mult}(r, 1/r)$ and the non-R-reparability of $F(r, \sqrt{r})$.

The generalization of the SK^- operators of section 2 is similarly obtained, by reversing the direction of the implication in the definition of R-reparability. We call this notion *anti-R-reparability* and define it as follows:

Definition 3.5. *Given a class of models* $\mathcal{M}$, *let* $c \in \bigcap_{M \in \mathcal{M}} D^M_{\mathsf{c}}$, *and suppose that every model where* $LD_{[\![F]\!]^M}([\![exp_1]\!]^{M^*}) = c$, *we have* $[\![exp_2]\!]^{M^*} = *$. *Then we say that the expression* $F(exp_1, exp_2)$ *is* **anti-R-repairable as** c.

In words: an expression $F(exp_1, exp_2)$ is anti-R-repairable when there is a value c of type c shared by all models, and in all models where c is L-determined by exp_1 for F, the evaluation of exp_2 fails.

The expression $F(r, \sqrt{r})$ of example 2 above is an instance of an anti-R-repairable expression, for the only value of r that L-determines the value of this expression, namely $r = -3$, fails the right-hand argument.

In the following definition we use the notion of anti-R-reparability to generalize the SK^- strategy.

Definition 3.6. *The (global)* $\textbf{SK}^-$ *strategy for* $[\![F(exp_1, exp_2)]\!]^{M^*}$ *is defined by:*

$$[\![F^{\text{SK}^-}(exp_1, exp_2)]\!]^{M^*} =$$

$$\begin{cases} [\![F(exp_1, exp_2)]\!]^M \\ \quad [\![exp_1]\!]^{M^*} \neq * \ and\ [\![exp_2]\!]^{M^*} \neq * \\ \\ c \quad LD_{[\![F]\!]^M}([\![exp_1]\!]^{M^*}) = c \in \bigcap_{M \in \mathcal{M}} D^M_{\mathsf{c}} \\ \quad and\ F(exp_1, exp_2)\ is\ R\text{-}repairable\ as\ c \\ \quad or\ anti\text{-}R\text{-}repairable\ as\ c \\ \\ * \quad otherwise \end{cases}$$

Definition 3.6 adds to definition 3.4 the possibility that the expression $F(exp_1, exp_2)$ is anti-R-repealable. The propositional SK^- operators of section 2 are instances of definition 3.6, which is also applicable to binary functions more generally. For instance, the expression $F(r, \sqrt{r})$ in example 2 satisfies $F^{\text{SK}^-}(r, \sqrt{r}) = 0$ when $r = -3$. Still, in terms of its repair potential, the SK^- strategy is weaker than the SK strategy. The following example illustrates that with a non-propositional expression.

Example 3: Let us consider the expression $G(r, \sqrt{s})$, where $G(x, y)$ is defined by 0 for $x < 0$ and by $x + y$ otherwise. In models where $r = -5$ and $s = -3$, the value of the left-hand argument

(-5) L-determines the value of G. Accordingly, when $r = -5$ and $s = -3$, the value of $G^{\text{SK}}(r, \sqrt{s})$ is 0 despite the failure of $\sqrt{s}$.[12] By contrast, there are models where the expression $\sqrt{s}$ fails and r is positive, hence does not L-determine the value of G. Conversely, there are also models where the expression $\sqrt{s}$ does not fail and r is negative, hence L-determines the value of G. This means that the expression $G(r, \sqrt{s})$ is neither R-repairable nor anti-R-repairable. As a result, in models where $r = -5$ and $s = -3$, the expression $G^{\text{SK}^-}(r, \sqrt{s})$ fails, unlike its SK parallel.

From the definitions above and examples 1-3 we conclude that the more general repair strategies for binary functions show the same hierarchy that we pointed out for the propositional connectives: the SK strategy is the most general repair strategy, WK allows no repair, whereas the repair strategies of WK^+ and SK^- are properly in between these two extremes.

4 Concluding remarks

This paper proposed new binary operators on truth-value denoting expressions, which, unlike the Weak Kleene connectives, allow filtering and conditional presuppositions, and unlike the Strong Kleene connectives, do not face the "proviso" problem. We defined asymmetric operators that allow left-to-right filtering (the "Weak Kleene plus" operators) as well as conditional presuppositions (the "Strong Kleene minus" operators) without proviso-like problems. These operators were generalized for arbitrary binary function expressions, which reveals the centrality of values that left/right-determine the result of a function for the treatment of presupposition projection in trivalent semantics.

One last note is in place about the special status of Strong Kleene (SK) connectives in the treatment of the third truth-value $*$. As has been previously observed (Muskens, 1995; Beaver and Krahmer, 2001), SK conjunction and disjunction are greatest lower bound and least upper bound operators, respectively, with respect to the less-or-equal partial order, where '1' and '0' are treated numerically and '$*$' is treated numerically as ½. This

---

[12]We may consider this kind of case as an illustration of a "proviso problem" for non-propositional binary functions: L-determination by the left-hand argument guarantees a successful evaluation of the function in cases when its right-hand argument fails, even if that failure is logically unrelated to L-determination.

gives the following lattice structure in SK trivalent logic (see Fitting (1991) for generalizations):

When it comes to theories of presupposition projection, this formal elegance has an empirical price: the SK-based partial order is a proper subset of the order determined by the Tarskian notion of entailment (Keenan, 1973; Beaver, 1997) in definition 1.1. Tarskian entailment in trivalent semantics supports the following preorder:

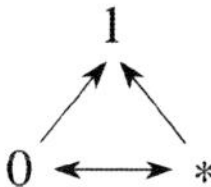

Importantly, when it comes to entailment in natural language there is reason to prefer the Tarskian preorder to the SK partial order. For instance, the sentence *Sue has stopped shouting* intuitively entails the sentence *Sue is not shouting*, but the former can denote '$*$' while the latter denotes '0' (e.g. in situations where Sue has just started shouting). This indicates that the preorder determined by the Tarskian condition is advantageous to the SK-based partial order as a basis for a trivalent semantics of presuppositions. The 0 and $*$ values are distinguished by their projection but no by their support of entailment relations. This is expected by the Tarskian preorder and not by the SK partial order. Thus, although the SK truth tables are logically natural, and indeed have led to interesting logical results, their modelling of the $*$ value as "unknown" or "in between 0 and 1" is the source of their proviso problem when used for meanings of natural language operators. A linguistically more adequate view ensues from treating the $*$ value as a "failure" or an "exception" as in the Weak Kleene connectives or (de Groote and Lebedeva, 2010). This requires further inquiries into intermediate systems like the WK^+ or SK^- operators that were studied above. These operators sacrifice truth-functionality – or, more generally, locality or at least extensionality – but model filtering and conditional presuppositions similarly to the Strong Kleene connectives, without running into their well-known problems.

Acknowledgments

I am grateful to three MOL reviewers, as well as to Lisa Bylinina, Danny Fox, Philippe de Groote, Ed Keenan, Matthew Mandelkern, Rick Nouwen, Jacopo Romoli and members of the presupposition reading group in Utrecht. Work on this paper was partially funded by the European Research Council (ERC) under the European Union's Horizon 2020 research and innovation programme (grant agreement No 742204).

References

David I. Beaver. 1997. Presupposition. In van Benthem and Alice ter Meulen, editors, *Handbook of Logic and Language*, pages 939–1008. Elsevier, Amsterdam.

David I. Beaver. 1999. Presupposition accommodation: A plea for common sense. In Lawrence S. Moss, Jonathan Ginzburg, and Maarten de Rijke, editors, *Logic, Language and Computation*, volume 2. CSLI Publications, Stanford, CA.

David I. Beaver. 2001. *Presupposition and Assertion in Dynamic Semantics*. CSLI Publications, Stanford, CA.

David I. Beaver and Bart Geurts. 2014. Presupposition. In Edward N. Zalta, editor, *The Stanford Encyclopedia of Philosophy*, winter 2014 edition.

David I. Beaver and Emiel Krahmer. 2001. A partial account of presupposition projection. *Journal of Logic, Language and Information*, 10(2):147.

Johan van Benthem. 1991. *Language in Action: Categories, Lambdas and Dynamic Logic*. North-Holland, Amsterdam.

Stephen Blamey. 1986. Partial logic. In D. Gabbay and F. Guenthner, editors, *Handbook of Philosophical Logic*, volume 3, pages 1–70. D. Reidel Publishing Company, Dordrecht.

Gennaro Chierchia and Sally McConnel-Ginet. 1990. *Meaning and Grammar: an introduction to semantics*. MIT Press, Cambridge, Mass.

Melvin Fitting. 1991. Kleene's logic, generalized. *Journal of Logic and Computation*, 1(6):797–810.

Gerald Gazdar. 1979. *Pragmatics: Implicature, Presupposition, and Logical Form*. Academic Press, New York.

Benjamin R. George. 2008. Presupposition repairs: a static, trivalent approach to predicting projection. Master thesis, UCLA.

Benjamin R. George. 2014. Some remarks on certain trivalent accounts of presupposition projection. *Journal of Applied Non-Classical Logics*, 24(1-2):86–117.

Bart Geurts. 1996. Local satisfaction guaranteed: A presupposition theory and its problems. *Linguistics and Philosophy*, 19(3):259–294.

Philippe de Groote and Ekaterina Lebedeva. 2010. Presupposition accommodation as exception handling. In *Proceedings of the 11th Annual Meeting of the Special Interest Group on Discourse and Dialogue*, pages 71–74. Association for Computational Linguistics.

Irene Heim. 1983. On the projection problem for presuppositions. In *Proceedings of WCCFL 2*, pages 114–125, Stanford, CA. CSLI Publications.

Lauri Karttunen and Stanley Peters. 1979. Conventional implicature. In Choon-Kyu Oh and David A. Dinneen, editors, *Syntax and semantics*, volume 11, pages 1–56. Academic Press, New York.

Edward L. Keenan. 1973. Presupposition in natural logic. *The Monist*, 57(3):344–370.

Daniel Lassiter. 2012. Presuppositions, provisos, and probability. *Semantics and Pragmatics*, 5:1–37.

Matthew Mandelkern. 2016. Dissatisfaction theory. In *Procedings of SALT 26*, pages 391–416.

Matthew Mandelkern, Jérémy Zehr, Jacopo Romoli, and Florian Schwarz. 2017. Asymmetry in presupposition projection: The case of conjunction. In *Procedings of SALT 27*, pages 504–524.

Reinhard Muskens. 1995. *Meaning and partiality*. CSLI Publications, Stanford, CA.

Stanley Peters. 1979. A truth-conditional formulation of Karttunen's account of presupposition. *Synthese*, 40(2):301–316.

Philippe Schlenker. 2011. The proviso problem: a note. *Natural Language Semantics*, 19(4):395–422.

Dependently-Typed Montague Semantics in the Proof Assistant Agda-flat

Colin Zwanziger
Department of Philosophy
Carnegie Mellon University
Pittsburgh, PA, USA
`zwanzig@cmu.edu`

Abstract

We apply the Agda-flat proof assistant (Vezzosi, 2019) to computational semantics. Computational semantics in Agda-flat is distinguished from the approach based on Coq (Chatzikyriakidis and Luo, 2014) in that it allows an implementation of the classical, intensional semantic analyses of Montague (1973). That is, it synthesizes the modern dependent type theory and Montague intensional logic traditions in the computational semantics setting. To demonstrate this, we show how to replicate Montague's analyses in the type theory of Zwanziger (2018), which closely corresponds to the Agda-flat system. Accompanying code type-checks these analyses in Agda-flat.

1 Introduction

Proof assistants, e.g. Coq (Chatzikyriakidis and Luo, 2014), have been applied in computational semantics to support natural language inference. By translating (declarative) sentences of a natural language into the language of a proof assistant, the problem of whether one natural language sentence implies another is reduced to finding a proof of implication in that proof assistant. Ideally, both the translation and the proof search are automated.

While proof assistants like Coq and the similar system Agda are adequate for implementing most semantic analyses given in the modern dependent type theory tradition (Ranta, 1994), they do not provide an obvious implementation of the classical intensional semantic analyses of Montague (1973). The criterion that a computational semantics system be general enough to capture Montague's analyses has been termed the "Montague Test" (Morrill and Valentín, 2016).

The present work introduces computational semantics in the Agda-flat proof assistant. Agda-flat (Vezzosi, 2019) is a variation on ordinary Agda implementing a so-called comonadic modal dependent type theory. Agda-flat was created by Andrea Vezzosi (c. 2017-2019) as a tool for verifying mathematical theorems in type-theoretical foundations. The logic behind the most recent mode of Agda-flat is essentially similar to **CHoTT**, which was isolated by Zwanziger (2018) as a natural dependently-typed, optionally hyperintensional analog of Montague's simply-typed intensional logic. As such, Agda-flat passes (the relevant incarnation of) the Montague Test. Furthermore, **CHoTT** can be used as a readable notation for Agda-flat.

Below, we develop enough of natural language semantics in **CHoTT** and Agda-flat to pass the Montague Test. In Section 2, we review the type theory **CHoTT** of Zwanziger (2018), which closely corresponds to the Agda-flat system, and is seen to provide a dependently-typed version of Montague intensional logic. In Section 3, the Montague Test sentences are translated into **CHoTT**. In Section 4, we discuss the Agda-flat system, and in particular the straightforward rendering of **CHoTT** sentences therein. Pursuant to the translations given in Section 3, an auxiliary repository is provided type-checking the translations of all Montague Test sentences.

2 Comonadic Homotopy Type Theory

We review the type theory **CHoTT** of Zwanziger (2018), which provides a dependently-typed version of the intensional logic **IL** of Montague (1973). To motivate **CHoTT**, a preview of the correspondence between **CHoTT** and **IL** is given in Subsection 2.1. Furthermore, some context for the development of **CHoTT** is provided in Subsection 2.2. The technical discussion of **CHoTT** is in Subsection 2.3.

IL	**CHoTT**	English Gloss
t	Prop	"type of truth values/propositions"
$\langle s, a \rangle$	$\flat A$	"type of intensions of terms of type a/A"
$\hat{\ }\alpha$	$t^\flat$	"the intension of term α/t"
$\check{\ }\alpha$	$t_\flat$	"the extension of term α/t"
$\Box\phi$	$\Box\phi$	"necessarily ϕ"

Table 1: An IL-to-CHoTT Dictionary

2.1 Cheat Sheet for CHoTT

As any extension of intensional logic should, the type theory **CHoTT** will include analogs of the intension, extension, and other operators of **IL**. These are foreshadowed in Table 1. The variant notation from **IL** follows Shulman (2018), and reflects the essentially independent origins of **IL** and Shulman's system.

Pursuant to the correspondence of Table 1, the translations of the Montague Test sentences given in Section 3 will bear an obvious similarity to those of Montague. There will be a few technical points of divergence between **CHoTT** and **IL**, though, discussed in the course of Subsection 2.3. Though these particularities of **CHoTT** can be justified on technical grounds, it is also worthwhile to understand **CHoTT** and its relation to **IL** in the context of the literature.

2.2 Provenance of CHoTT

We here review some historical context and justification for **CHoTT**. A reader less concerned by such matters can skip ahead to Subsection 2.3 for the technical development.

The type theory **CHoTT** is a "comonadic homotopy type theory", combining a version of Montague's intensional logic ("comonadic type theory") with homotopy type theory (HoTT), a version of dependent type theory. We visit each of these aspects in turn.

2.2.1 Intensional Logic and Comonadic Type Theory

The identification of Montague intensional logic with comonadic type theory dates to joint work of the author (Awodey et al., 2015), and is developed in the M.S. thesis of the author (Zwanziger, 2017), in both syntactic and semantic aspects. Comonadic type theory is a variety of modal type theory which was first treated systematically in the 1990's, with approaches due to Bierman and de Paiva (2000) and Davies and Pfenning (Pfen-

ning and Davies, 2001). Montague's intensional logic can thus be understood retrospectively as the original comonadic type theory. However, Montague's syntax lacks β-conversion (AKA λ-conversion) and the usual rules for quantifiers, an issue addressable within these modern approaches (Zwanziger, 2017). The present paper follows the Davies and Pfenning approach, and and in particular Shulman (2018)'s extension to homotopy type theory.

2.2.2 Homotopy Type Theory

HoTT (Univalent Foundations Program, 2013) is a proposed type-theoretic alternative to axiomatic set theory as a foundation for math. HoTT incorporates dependent type theory, a framework which has been fruitfully applied in natural language semantics since Sundholm (1989) and Ranta (1994). Recently, dependent type theory has been called "modern type theory" in the context of natural language semantics, to distinguish it from the simply-typed intensional logic of Montague (Chatzikyriakidis and Luo, 2018). Particularly appealing applications of dependent type theory in natural languages semantics, dating to the early work, include the use of the baked-in bounded quantification to model the bounded quantification of natural language (part of the "common nouns-as-types" viewpoint) (Luo, 2012), and a natural approach to dynamic semantics.

As for the "homotopy" part, it is argued in Zwanziger (2018) that this provides a natural approach to hyperintensional natural language semantics. As a homotopy type theory, **CHoTT** includes two notions of equality: definitional equality, written $\equiv$, and thought of as (hyper)intensional equality, and typal equality, written $=$, and thought of as an *a posteriori* equality. In line with this intuition, $\equiv$ is stronger than $=$. In **CHoTT**, we find 'intensional' operators which respect only $\equiv$, not $=$. Furthermore, this more "granular" intensional equality need not conflate statements with

$$\frac{}{\cdot \mid \cdot \ \mathsf{ctx}} \ \text{ctx-Emp.}$$

$$\frac{\Delta \mid \Gamma \vdash B : U}{\Delta \mid \Gamma, x : B \ \mathsf{ctx}} \ \text{ctx-Ext.}^{e} \qquad\qquad \frac{\Delta \mid \Gamma, x : A, \Gamma' \ \mathsf{ctx}}{\Delta \mid \Gamma, x : A, \Gamma' \vdash x : A} \ \text{Var.}^{e}$$

$$\frac{\Delta \mid \cdot \vdash B : U}{\Delta, u : \{\flat\} \ B \mid \cdot \ \mathsf{ctx}} \ \text{ctx-Ext.}^{i} \qquad\qquad \frac{\Delta, u : \{\flat\} \ A, \Delta' \mid \Gamma \ \mathsf{ctx}}{\Delta, u : \{\flat\} \ A, \Delta' \mid \Gamma \vdash u : A} \ \text{Var.}^{i}$$

Table 2: The Extensional $(-^{e})$ and Intensional $(-^{i})$ Context Rules

the same truth conditions. A hyperintensional semantics of **CHoTT**, inspired by the homotopy-theoretic semantics of HoTT, is the subject of other research by the author (Zwanziger, 2019).

In summary, **CHoTT** and the closely related Agda-flat, while building directly on a separate tradition than intensional logic, do incorporate intensional logic, together with additional features. Furthermore, these extra features have various uses for the natural language semanticist, rather than being peculiarities inflicted on them by the Agda-flat system. That said, since our intention is to make a beeline for formalizing the Montague Test sentences, these modeling advantages will not be pursued below, with the exception of common nouns-as-types.

2.3 The System CHoTT

We now turn to delineating the system **CHoTT**, with a focus on the similarities and differences with **IL**.

Officially, **CHoTT** is the fragment of Shulman (2018) consisting of the usual notions of homotopy type theory, together with the comonadic type operator $\flat$, which we think of as an intension type operator performing the role of Montague's $\langle s, - \rangle$. In less condensed terms, we have the following:

In the tradition of Pfenning and Davies (2001) and Shulman (2018), **CHoTT** has two variable judgements,

$$u : \{\flat\} \ A$$

and

$$x : A$$

We will say (at variance with prior terminology) that "u is an intensional variable of type A," when $u : \{\flat\} \ A$ and that "x is an extensional variable of type A," when $x : A$.

Remark 1 *Intuitively, having an assumption $u : \{\flat\} \ A$ of an intensional variable of type A is akin to having an assumption $x : \flat A$ of an extensional variable of the type of intensions of terms of type A. Indeed, such assumptions will turn out to be interchangeable.*

The hypothetical judgements of **CHoTT** have the form

$$\Delta \mid \Gamma \vdash t : B$$

and

$$\Delta \mid \Gamma \vdash t \equiv u : B$$

where Δ represents a list of intensional typed variables, and Γ a list of extensional typed variables. This $\Delta \mid \Gamma$ is called the context, and we will have, as usual, that a type appearing in the context may depend only on typed variables to its left. So types in Γ can depend on variables in Δ, but not vice versa.

Due to the two variable judgements, there is a duplication of the rules for context extension and variables, with variants for extensional and intensional variables. These are given in Table 2. In view of Remark 1, the rule Var.^{i} is understood as the principle that an assumption of an intension of a term of type A yields a term of type A. This rule is, naturally, implicated in the derivation of the Montague extension operator, below.

We import the usual homotopy-type-theoretical notions (Univalent Foundations Program, 2013), including $\prod$- and $\sum$-types (analogs of the quantifiers $\forall$ and $\exists$), type universes and universe polymorphism, $=$-types, higher inductive types (HITs), and univalence. However, as a simplifying assumption, the typing rules are assumed to manipulate extensional variables only. For instance, the formation rule for $\prod$ is

$$\frac{\Delta \mid \Gamma \vdash A : U \qquad \Delta \mid \Gamma, x : A \vdash B : U}{\Delta \mid \Gamma \vdash \prod_{x:A} B : U}$$

$$\frac{\Delta \mid \cdot \vdash B : U}{\Delta \mid \Gamma \vdash \flat B : U} \; \flat\text{-Form. (Montague's } \langle s, - \rangle) \qquad \frac{\Delta \mid \cdot \vdash t : B}{\Delta \mid \Gamma \vdash t^{\flat} : \flat B} \; \flat\text{-Intro. (Montague's } {}^{\wedge}(-))$$

$$\frac{\Delta \mid \Gamma, x : \flat A \vdash B : U \qquad \Delta \mid \Gamma \vdash s : \flat A \qquad \Delta, u : \{\flat\} A \mid \Gamma \vdash t : B[u^{\flat}/x]}{\Delta \mid \Gamma \vdash (\text{let } u^{\flat} := s \text{ in } t) : B[s/x]} \; \flat\text{-Elim.}$$

$$\frac{\Delta \mid \Gamma, x : \flat A \vdash B : U \qquad \Delta \mid \cdot \vdash s : A \qquad \Delta, u : \{\flat\} A \mid \Gamma \vdash t : B[u^{\flat}/x]}{\Delta \mid \Gamma \vdash \text{let } u^{\flat} := s^{\flat} \text{ in } t \equiv t[s/u] : B[s^{\flat}/x]} \; \flat\text{-}\beta\text{-Conversion}$$

Table 3: The Rules for $\flat$

$$\frac{\Delta \mid \cdot \vdash B : U}{\Delta \mid \Gamma, x : \flat B \vdash B : U} \; \text{Weakening} \qquad \frac{\Delta \mid \Gamma \vdash t : \flat B \qquad \overline{\Delta, u : \{\flat\} B \mid \Gamma \vdash u : B} \; \text{Var.}^{i}}{\Delta \mid \Gamma \vdash (\text{let } u^{\flat} := t \text{ in } u) \equiv: t_{\flat} : B} \; \flat\text{-Elim.}$$

Figure 1: Derivation of $\flat$-Elim.-Simp. (Montague's ${}^{\vee}(-))$

in which $x : A$ is required to be extensional.

Finally, we have a comonad $\flat$ corresponding to Montague's $\langle s, - \rangle$, the rules for which appear as Table 3.

Note that the $\flat$-Form. and $\flat$-Intro. (intension operator) rules only apply "in an intensional context", that is, when no extensional variables are present in the context.[1] Thus, in a well-typed term, any intensional operator ($\flat(-)$ or $(-)^{\flat}$) that appears must have been adduced during a phase of the derivation with an intensional context.[2] Since variables within the scope of an intensional operator must receive *de re* interpretation (Zwanziger, 2017), and only intensional variables occur within the scope of an intensional operator, intensional variables can be motivated as a technical device for keeping track of which locations within a term receive *de re* interpretation.

The $\flat$-Elim. rule provides a way of substituting a term of type $\flat A$ for an intensional variable of type A via an explicit let-notation. This principle is in keeping with Remark 1. Furthermore, it gives a benign way of "substituting into an intensional context". To illustrate, such a substitution

$$\text{let } u^{\flat} := s \text{ in } t(u^{\flat})^{\flat} \quad ,$$

in which s receives *de re* interpretation, does not in general $\flat$-β-reduce to the *de dicto* form

$$t(s)^{\flat} \quad ,$$

and thus avoids the usual pitfall of substitution into intensional contexts. As a corollary, β-conversion (for function types) clearly gives

$$(\lambda x.\text{let } u^{\flat} := x \text{ in } t(u^{\flat})^{\flat})(s) \equiv$$
$$\text{let } u^{\flat} := s \text{ in } t(u^{\flat})^{\flat} \quad ,$$

not

$$(\lambda x.\text{let } u^{\flat} := x \text{ in } t(u^{\flat})^{\flat})(s) \equiv t(s)^{\flat} \quad ,$$

explaining why β-conversion is benign in **CHoTT**.

Though $\flat$-Elim. is subtle, the reader may take heart that we can use it to derive an 'extension' operator corresponding to Montague's ${}^{\vee}(-)$, which, again following Shulman (2018), we call $(-)_{\flat}$. That is, the rule

$$\frac{\Delta \mid \Gamma \vdash t : \flat B}{\Delta \mid \Gamma \vdash t_{\flat} : B} \; \flat\text{-Elim.-Simple}$$

is derivable (Figure 1). As foreshadowed above, this derivation also makes crucial use of the rule Var.^{i}.

Furthermore, we have the conversion $(t^{\flat})_{\flat} \equiv t$, familiar from Montague. That is, the principle

$$\frac{\Delta \mid \cdot \vdash t : B}{\Delta \mid \Gamma \vdash (t^{\flat})_{\flat} \equiv t : B} \; \flat\text{-}\beta\text{-Conv.-Simp.}$$

---

[1] While **IL** does not place any similar restriction on the intension operator ${}^{\wedge}(-)$, Montague does stipulate a constant-domain Kripke semantics for **IL**. The analogous Kripke-Montague semantics for **CHoTT** need not stipulate constant domains. But a type $\flat A$ of intensions will, naturally, have a constant domain interpretation regardless. Remark 1 relates extensional variables of type $\flat A$ with intensional variables of type A.

[2] In the current terminology, then, an intensional context is not, as traditionally, *defined* by the presence of an intensional operator, but rather a more primitive notion that is prerequisite for the presence of an intensional operator.

43

is derivable, obtained with

$$(t^\flat)_\flat \equiv \mathsf{let}\ u^\flat := t^\flat\ \mathsf{in}\ u \quad (\text{Def'n. of } (-)_\flat)$$
$$\equiv t \quad\quad\quad (\flat\text{-}\beta\text{-Conv.})$$

as calculation.

It is with all the comonadic rules in concert that Remark 1 is ultimately justified. The reader may confirm, for instance, that the principles

$$\frac{u : \{\flat\}\ A \mid \cdot \vdash t : B}{\cdot \mid x : \flat A \vdash \mathsf{let}\ u^\flat := x\ \mathsf{in}\ t : \mathsf{let}\ u^\flat := x\ \mathsf{in}\ B}$$

and

$$\frac{\cdot \mid x : \flat A \vdash t : B}{u : \{\flat\}\ A \mid \cdot \vdash t[u^\flat/x] : B[u^\flat/x]}$$

are derivable.

3 Natural Language Semantics in CHoTT

We now set about rendering the Montague Test sentences in **CHoTT**.

To illustrate the issues involved, let's examine what will be the translations of the sentence, "John believes that a fish walks." This sentence is traditionally held to have two construals, a *de dicto* and a *de re*. These construals are to be represented by

$$believe(j, (\exists(x : fish).walk(i(x)))^\flat)$$

and

$$\exists(x : fish).\mathsf{let}\ u^\flat := i(x)\ \mathsf{in}$$
$$believe(j, walk(u^\flat)^\flat)\quad ,$$

respectively. These formulas exhibit several less familiar features which bear discussion, including defined (non-primitive) notation for logic inside HoTT ($\exists$), common nouns-as-types ($x : fish$) and associated type coercions (i), and the handling of *de re* with (let-substitutions for) intensional variables.

In Subsection 3.1, we explicate any prerequisite defined notation and associated theory. In Subsection 3.2, we give the **CHoTT** renderings of the Montague Test sentences.

3.1 Defined Notations

We now develop several notions for eventual use in Subsection 3.2. Unless otherwise specified, the definitions of the present subsection are adapted from Chapter 3 of the HoTT Book (Univalent Foundations Program, 2013).

3.1.1 Predicate Logic

Central to the approach of the current section is the use of the type Prop of "propositions" in lieu of Montague's type t of "truth values". Within HoTT, propositions are identified with those types that have at most one term (up to the equality $=$). Intuitively, such types encode no more (extensional) information than whether they have a term ("are inhabited") or not. In such case, truth is identified with inhabitation. Formally speaking, we have the following definitions:

Definition 2 *For any type* $A \ :\ U$, *let* $\mathrm{isProp}(A) :\equiv \prod_{x,y:A} x = y$.

When $\mathrm{isProp}(A)$ is inhabited, we say that A is a proposition, and when such A is inhabited, we say A is true.

Definition 3 $\mathrm{Prop} :\equiv \sum_{A:U} \mathrm{isProp}(A)$

Whereas a sequent of form $\Delta \mid \Gamma, x\ :\ A \vdash B\ :\ U$ is thought of as a type depending on A, a sequent of form $\Delta \mid \Gamma, x\ :\ A \vdash P\ :\ \mathrm{Prop}$ may be thought of as a predicate on A, rather than a "proposition depending on A". This notion of predicate gives rise to a notion of predicate logic inside type theory with the following constructs:

Definition 4

$$\top :\equiv 1$$
$$\bot :\equiv 0$$
$$P \wedge Q :\equiv P \times Q$$
$$P \vee Q :\equiv \|P + Q\|$$
$$P \Rightarrow Q :\equiv P \to Q$$
$$\neg P :\equiv P \Rightarrow \bot$$
$$\forall(x : A)P(x) :\equiv \prod_{x:A} P(x)$$
$$\exists(x : A)P(x) :\equiv \left\|\sum_{x:A} P(x)\right\|\quad ,$$

where $\Delta \mid \Gamma, x : A \vdash P, Q : \mathrm{Prop}$.

Here, $\|-\|$ denotes "propositional truncation", as defined in Chapter 3 of the HoTT Book (Univalent Foundations Program, 2013), which quotients any type to a proposition.

To the logic of the HoTT Book, we do add one definition:

Definition 5 *Let* $\Delta, u : \{\flat\}\ A \mid \cdot \vdash P : \mathrm{Prop}$. *Then* $\Box P :\equiv \|\flat P\|$.

This pregnant and perhaps surprising definition can be understood in terms of the type-theoretic

$$E : U$$
$$j, b : \flat E$$
$$walk, man, talk, fish, woman, unicorn, park : \flat E \to \text{Prop}$$
$$believe : \flat\, \text{Prop} \to \flat E \to \text{Prop}$$
$$seek, catch, eat, find, love, lose : \flat(\flat(\flat E \to \text{Prop}) \to \text{Prop}) \to \flat E \to \text{Prop}$$
$$slowly, try : \flat(\flat E \to \text{Prop}) \to \flat E \to \text{Prop}$$
$$in : \flat(\flat(\flat E \to \text{Prop}) \to \text{Prop}) \to \flat(\flat E \to \text{Prop}) \to$$
$$\flat E \to \text{Prop}$$

Figure 2: Lexicon for the Montague Test

tendency to conflate truth with inhabitation. In Kripke-Montague semantics, intensions for (the interpretation of) a type A only exist when (the domain interpreting) A at each world is non-empty. So $\flat A$ at a world is non-empty (*i.e.* true) when A is nonempty (*i.e.* true) at every world.[3]

3.1.2 Subtypes and Coercions

We will have reason to view common nouns as types (for the purposes of type-bounded quantification). Yet predicates like "talks" are apt to apply to women, men, donkeys (perhaps), *etc.* Consequently, "talks" is better interpreted as a predicate on a type of entities subsuming women, men, and donkeys. To accommodate both impulses, we require a notion of subtype and type coercion, which will allow us to write logical forms like, *e.g.*, $\forall(x : man).talk(i(x))$ for "Every man talks." This motivates our use of the subtypes of HoTT, defined as follows:

Definition 6 *Let* $\Delta \mid \Gamma, x : A \vdash P : \text{Prop}$. *Then* $\{x : A \mid P(x)\} :\equiv \sum_{x:A} P(x)$. *Furthermore let* $i :\equiv \pi_1 : \sum_{x:A} P(x) \to A$.

This $\{x : A \mid P(x)\}$ is said to be the subtype of A satisfying property P, and i the inclusion or coercion of $\{x : A \mid P(x)\}$ into A.

For readability, we will furthermore coin the following abuse of notation:

Convention 7 *The type* $\{x : A \mid P(x)\}$, *where* $\Delta \mid \Gamma, x : A \vdash P : \text{Prop}$, *may simply be denoted by* P, *where confusion is unlikely.*

The formula $\forall(x : man).talk(i(x))$ is now understood as using Convention 7, as well as a coercion.

3.2 The Montague Test Sentences

With the benefit of the foregoing, we now render the Montague Test sentences in **CHoTT**. We will treat exactly the sentence suite formulated by Morrill and Valentín (2016) in introducing the Montague Test.

We assume a ground type E, together with a number of constants that serve as the translations of natural language terms, detailed in Figure 2. These are simply taken from Dowty *et al.* (1981), Chapter 7, using the correspondences suggested by Table 1, plus a single important stipulation: wherever the type e appears for Dowty *et al.*, $\flat E$ (rather than E) appears for us. This unfortunate technical device is necessitated by our approach to *de re* readings, which makes use of let-substitution. One can substitute terms of type $\flat E$ using let, but not E.[4]

The **CHoTT** forms for the Montague Test sentences are given in Table 4, below. Any reference numbers come from Chapter 7 of Dowty *et al.* (1981).

Overall, **CHoTT** formulas of Table 4 closely reflect those given by Montague (1973). To illustrate the similarities and differences, we return, at last, to the translations of "John believes that a fish walks." The *de dicto* reading is translated as

$$believe(j, (\exists(x : fish).walk(i(x)))^{\flat}) \quad .$$

---

[3]It would be more satisfying to have simply $\Box P :\equiv \flat P$, but **CHoTT** is apparently too weak to prove that $\flat P$ is a proposition, as desired of $\Box P$. This gives a hint that **CHoTT**, despite its advantages, may not be the last word in comonadic type theory.

[4]One way around this problem would be to use adjoint type theory (Benton and Wadler, 1996; Licata and Shulman, 2016), in which some ground types already have "constant domain", in lieu of comonadic type theory.

Ref. No.	Sentence	Translation(s)
7	John walks.	$walk(j)$
16	Every man talks.	$\forall(x : man).talk(i(x))$
19	The fish walks.	$\exists(x : fish).walk(i(x)) \wedge \forall(y : fish).\|y = x\|$
32	Every man walks or talks.	$\forall(x : man).walk(i(x)) \vee talk(i(x))$
34	Every man walks or every man talks.	$(\forall(x : man).walk(i(x))) \vee (\forall(x : man).talk(i(x)))$
39	A woman walks and she talks.	$\exists(x : woman).walk(i(x)) \wedge talk(i(x))$
43	John believes that a fish walks.	$believe(j, (\exists(x : fish).walk(i(x)))^{\flat})$
45		$\exists(x : fish).\mathsf{let}\ u^{\flat} := i(x)\ \mathsf{in}\ believe(j, walk(u^{\flat})^{\flat})$
48	Every man believes that a fish walks.	$\exists(x : fish).\mathsf{let}\ u^{\flat} := i_1(x)\ \mathsf{in}$ $\forall(y : man).believe(i_2(y), walk(u^{\flat})^{\flat})$
49		$\forall(y : man).\exists(x : fish).\mathsf{let}\ u^{\flat} := i_1(x)\ \mathsf{in}$ $believe(i_2(y), walk(u^{\flat})^{\flat})$
n/a		$\forall(y : man).believe(i_2(y), (\exists(x : fish).walk(i_1(x)))^{\flat})$
57	Every fish such that it walks talks.	$\forall(x : fish \wedge walk).talk(i(x))$
60	John seeks a unicorn.	$seek(j, (\lambda(P : \flat(\flat E \to \mathrm{Prop})).\exists(x : unicorn).P_{\flat}(i(x)))^{\flat})$
62		$\exists(x : unicorn).\mathsf{let}\ u^{\flat} := i(x)\ \mathsf{in}$ $seek(j, (\lambda(P : \flat(\flat E \to \mathrm{Prop})).P_{\flat}(u^{\flat}))^{\flat})$
73	John is Bill.	$\|j = b\|$
76	John is a man.	$man(j)$
83	Necessarily, John walks.	$\Box(walk(j))$
86	John walks slowly.	$slowly(walk^{\flat})(j)$
91	John tries to walk.	$try(j, walk^{\flat})$
94	John tries to catch a fish and eat it.	$try(j, (\lambda(y : \flat E).\mathsf{let}\ u^{\flat} := y\ \mathsf{in}\ \exists(x : fish).$ $catch'((\lambda(P : \flat(\flat E \to \mathrm{Prop})).P_{\flat}(u^{\flat}))^{\flat})(i(x)) \wedge$ $eat'((\lambda(P : \flat(\flat E \to \mathrm{Prop})).P_{\flat}(u^{\flat}))^{\flat})(i(x)))^{\flat})$
98	John finds a unicorn.	$\exists(x : unicorn).\mathsf{let}\ u^{\flat} := i(x)\ \mathsf{in}$ $find((\lambda(P : \flat(\flat E \to \mathrm{Prop})).P_{\flat}(u^{\flat}))^{\flat})(j)$
105	Every man such that he loves a woman loses her.	$\exists(y : woman).\mathsf{let}\ u^{\flat} := i_2(y)\ \mathsf{in}$ $\forall(x : man \wedge love((\lambda(P : \flat(\flat E \to \mathrm{Prop})).P_{\flat}(u^{\flat}))^{\flat})).$ $lose((\lambda(P : \flat(\flat E \to \mathrm{Prop})).P_{\flat}(u^{\flat}))^{\flat})(i_1(x))$
110	John walks in a park.	$\exists(x : park).\mathsf{let}\ u^{\flat} := i(x)\ \mathsf{in}$ $in((\lambda(P : \flat(\flat E \to \mathrm{Prop})).P_{\flat}(u^{\flat}))^{\flat})(walk^{\flat})(j)$
116	Every man doesn't walk.	$\neg\forall(x : man).walk(i(x))$
118		$\forall(x : man).\neg walk(i(x))$

Table 4: Montague Test Sentences

Aside from being pleasingly compact, in keeping with dependent type theory's type-bounded quantification, this formula is hardly changed from Montague (1973).

Things get more interesting for the *de re* reading, which is represented by

$$\exists(x : fish).\mathsf{let}\ u^{\flat} := i(x)\ \mathsf{in}$$
$$believe(j, walk(u^{\flat})^{\flat})\quad,$$

making use of the let-substitution not familiar from **IL**. To explicate, it is perhaps helpful to think of $walk(u^{\flat})$ as $walk(y)[u^{\flat}/y]$. That is, $walk(y)$ was altered to have an intensional variable, in keeping with Remark 1. As discussed in Section 2.3, the intensional variable u signals *de re* interpretation, and makes it possible to adduce an intension operator, forming $walk(u^{\flat})^{\flat}$. We later desire a(n extensional) quantification, so we switch back to extensional variables by substituting into the intensional context, forming

$$\mathsf{let}\ u^{\flat} := i(x)\ \mathsf{in}\ believe(j, walk(u^{\flat})^{\flat})\quad,$$

and then quantify.

In sum, this example, as well as the rest of the Montague Test sentences, are well-handled in **CHoTT**, and manifestly similarly to in **IL**.

4 Agda-flat

We now discuss Agda-flat, and the implementation therein of the Montague Test sentences, as rendered in **CHoTT** in the previous section.

4.1 About Agda-flat

Agda-flat (Vezzosi, 2019) is a proof assistant due to Andrea Vezzosi (c. 2017-2019). It is a branch of the Agda proof assistant (Norell, 2007) which grew out of Vezzosi's work on modal dependent type theory (Nuyts et al., 2017). Agda is similar to other dependently-typed proof assistants, like Coq. Agda-flat adds on a modal operator, $\flat$.

Agda-flat was intended as a proof assistant for mathematical theorems. Early mathematical uses include Licata *et al.* (2018) and Wellen (2018). The original version of Agda-flat is largely incompatible with linguistic applications, due to its relatively strong proof theory. Amongst other things, one can prove that $\flat$ is idempotent (that is, $\flat\flat A \simeq \flat A$), which does not hold in the familiar model of Montague (1973). Following a 2019 request from the author, Vezzosi implemented a new mode, called ("no-flat-split"), with a weaker proof theory. It is this "no-flat-split" mode which corresponds closely to **CHoTT**, and thus provides a setting for (optionally hyperintensional) Montague semantics.

4.2 Agda-flat for Computational Semantics

4.2.1 Installation and Configuration

Agda-flat is available for download at: `https://github.com/agda/agda/tree/flat`. Unfortunately, the process of installing Agda and turning on Agda-flat is system-dependent and potentially complex. The Agda Wiki (Agda Collaboration) can provide guidance, and one may wish to avail oneself of a friendly Agda user, including online.

As mentioned in Section 4.1, the "no-flat-split" option is necessary for computational semantics applications. To engage the "no-flat-split" option, place the code

```
{-# OPTIONS --no-flat-split #-}
```

at the beginning of a module.

4.2.2 Use

Agda-flat and **CHoTT** are similar, with both admitting a $\flat$-operator, and in particular having a way of restricting the $\flat$-Form. and $\flat$-Intro operations. Whereas in **CHoTT**, this is achieved by restricting these operations to an intensional context, in Agda, the variable contexts are not made explicit, and instead heavy use is made of $\prod$-types. Of necessity, then, the $\prod$-types of Agda-flat are permitted to be of the form $\prod_{u:\{\flat\}\ A} P$, in which an intensional variable is bound. In "Agda notation", however, $\prod_{u:\{\flat\}\ A} P$ is written $(u\ :\{\flat\}\ A)\ \rightarrow\ P$.

Agda-flat allows a similar manipulation of intensional variables by the rules for inductive types. Furthermore, $\flat$ is actually defined as an inductive type in Agda-flat, and uses pattern matching instead of an Elim. rule, as for any inductive types in Agda.

That is, $\flat$ is defined as an inductive type by, in pseudo-code,

```
data ♭ (A :{♭} Type) :    Type where
   int :   (a :{♭} A) → ♭A          .
```

Whenever the $\flat$-Elim. rule would be used in **CHoTT**, we instead use pattern matching. For instance, we define Montague's extension operator thus:

```
ext:  {A :{♭} Type} → (♭A → A)
ext (int u) = u               .
```

Once the $\flat$ operator is defined, and the above stylistic differences between **CHoTT** and Agda-flat accounted for, the rendering of **CHoTT** terms in Agda-flat proceeds as expected.

4.2.3 The Montague Test

The logical forms given in Table 4 are rendered in type-checkable code, available on the GitHub of the author at `https://github.com/zwanzigerc/Montague-Test`. This repository makes use of the conventions and codebase of the HoTT-Agda library (Brunerie et al.). In addition to the "official" renderings, there are also alternative ones which use the universe `Type` in lieu of the more complex `Prop`.

Agda-flat thus satisfies the Montague Test, and is seen to accommodate computational dependently typed Montague semantics.

5   Future Work

With the basic setup of intensional computational semantics in Agda-flat achieved, a key goal becomes automation, both of the translation from natural language into Agda-flat, and of proof search. Since Agda is so similar to Coq, it is possible that Agda-flat could be integrated with the translation from Grammatical Framework used in FraCoq (Bernardy and Chatzikyriakidis, 2017). As for proof search, Agda has some limited automation options, such as the command "auto", which should be explored. Ultimately the performance of Agda-flat at natural language inference should be compared to other systems that satisfy the Montague Test, such as CatLog3 (Morrill, 2017).

In summary, **CHoTT** and the closely related Agda-flat, while building directly on a separate tradition than intensional logic, do incorporate intensional logic, together with additional features. Furthermore, these extra features have various uses for the natural language semanticist, rather than being peculiarities inflicted on them by the Agda-flat system. That said, since our intention is to m

Agda-flat (without the assumption of intensional $=$-induction) is furthermore an appropriate system to implement *hyper*intensional semantics in the style of Zwanziger (2018). A full discussion of this is deferred to later work.

Acknowledgments

This paper was made possible by Andrea Vezzosi's programming of Agda-flat. In particular, many thanks are due for the new option, "no-flat-split", which makes the system general enough for the purposes of the present paper. Felix Wellen and Zesen Qian provided valuable help with Agda. And thank you to Steve Awodey for suggesting and supporting my focus on Agda-flat.

The logical forms given for purposes of the Montague Test (which in particular do not use common noun types to model verb selectional restrictions) were influenced by personal communication with Carl Pollard.

Any errors in the present work are my own.

References

The Agda Collaboration. The Agda wiki. `https://wiki.portal.chalmers.se/agda/pmwiki.php`. Accessed: 2019-04-22.

Steve Awodey, Ulrik Buchholtz, and Colin Zwanziger. 2015. Comonadic categorical semantics of montague's intensional logic. In *Slides from Dynamics Semantics Workshop: Modern Type Theoretic and Category Theoretic Approaches*, Ohio State University, Columbus, Ohio.

Nick Benton and Philip Wadler. 1996. Linear logic, monads and the lambda calculus. In *Proceedings 11th Annual IEEE Symposium on Logic in Computer Science*, pages 420–431. IEEE.

Jean-Philippe Bernardy and Stergios Chatzikyriakidis. 2017. A type-theoretical system for the FraCaS test suite: Grammatical framework meets Coq. In *Long Papers of the 12th International Conference on on Computational Semantics (IWCS)*.

Gavin Bierman and Valeria de Paiva. 2000. On an intuitionistic modal logic. *Studia Logica*, 65(3):383–416.

Guillaume Brunerie, Kuen-Bang Hou (Favonia), Evan Cavallo, Tim Baumann, Eric Finster, Jesper Cockx, Christian Sattler, Chris Jeris, Michael Shulman, et al. Homotopy type theory in Agda. `https://github.com/HoTT/HoTT-Agda`. Accessed: 2019-04-22.

Stergios Chatzikyriakidis and Zhaohui Luo. 2014. Natural language inference in Coq. *Journal of Logic, Language and Information*, 23(4):441–480.

Stergios Chatzikyriakidis and Zhaohui Luo. 2018. *Formal Semantics in Modern Type Theories*. Wiley and ISTE Science Publishing Ltd.

David R. Dowty, Robert Wall, and Stanley Peters. 1981. *Introduction to Montague Semantics*. Springer Science and Business Media.

Daniel R. Licata, Ian Orton, Andrew M. Pitts, and Bas Spitters. 2018. Internal universes in models of homotopy type theory. *arXiv preprint arXiv:1801.07664*.

Daniel R. Licata and Michael Shulman. 2016. Adjoint logic with a 2-category of modes. In *International Symposium on Logical Foundations of Computer Science*, pages 219–235. Springer.

Zhaohui Luo. 2012. Common nouns as types. In *Proceedings of the International Conference on Logical Aspects of Computational Linguistics*, pages 173–185. Springer.

Richard Montague. 1973. The proper treatment of quantification in ordinary English. In K. Hintikka, J. Moravcsik, and Suppes P., editors, *Approaches to Natural Language*, pages 221–242. D. Reidel, Dordrecht.

Glyn Morrill. 2017. Parsing logical grammar: Cat-Log3. In *Proceedings of the Workshop on Logic and Algorithms in Computational Linguistics 2017*, pages 107–131, Stockholm. Stockholm University.

Glyn Morrill and José Oriol Valentín. 2016. Computational coverage of type logical grammar: The Montague Test. In *Empirical Issues in Syntax and Semantics 11*, pages 141–170.

Ulf Norell. 2007. Towards a practical programming language based on dependent type theory. Ph.D. thesis, Chalmers University of Technology.

Andreas Nuyts, Andrea Vezzosi, and Dominique Devriese. 2017. Parametric quantifiers for dependent type theory. In *Proceedings of the ACM on Programming Languages, ICFP*. ACM.

Frank Pfenning and Rowan Davies. 2001. A judgmental reconstruction of modal logic. *Mathematical Structures in Computer Science*, 11(4):511–540.

Aarne Ranta. 1994. *Type-Theoretical Grammar*. Oxford University Press.

Michael Shulman. 2018. Brouwer's fixed-point theorem in real-cohesive homotopy type theory. *Mathematical Structures in Computer Science*, 28(6):856–941.

Göran Sundholm. 1989. Constructive generalized quantifiers. *Synthese*, 79(1):1–12.

The Univalent Foundations Program. 2013. *Homotopy Type Theory: Univalent Foundations of Mathematics*. `https://homotopytypetheory.org/book`, Institute for Advanced Study, Princeton.

Andrea Vezzosi. 2019. Agda-flat. `https://github.com/agda/agda/tree/flat`. Accessed: 2019-04-22.

Felix Wellen. 2018. Flat. `https://github.com/felixwellen/DCHoTT-Agda/blob/master/Flat.agda`. Code library. Accessed: 2019-04-22.

Colin Zwanziger. 2017. Montague's intensional logic as comonadic type theory. M.S. thesis, Carnegie Mellon University, Pittsburgh.

Colin Zwanziger. 2018. Propositional attitude operators in homotopy type theory: Extended abstract. `https://colinzwanziger.com/wp-content/uploads/2019/06/nlcs-2018-extended-abstract.pdf`. In *Fifth Workshop on Natural Language and Computer Science*, Oxford.

Colin Zwanziger. 2019. Natural model semantics for comonadic and adjoint modal type theory. In *Proceedings of the Second Applied Category Theory Conference*, Compositonality.

Quantifier-free Least Fixed Point Functions for Phonology

Jane Chandlee
Haverford College
Haverford, PA, USA
jchandlee@haverford.edu

Adam Jardine
Rutgers University
New Brunswick, NJ, USA
adam.jardine@rutgers.edu

Abstract

In this paper we define *quantifier-free least fixed point functions (QFLFP)* and argue that they are an appropriate and valuable approach to modeling phonological processes (construed as *input-output maps*). These functions, characterized in terms of first order logic interpretations over graphs, provide a close fit to the observed typology, capturing both local and long-distance phenomena, but are also restrictive in desirable ways. Namely, QFLFP logical functions approximate the computation of deterministic finite-state transducers, which have been argued to form a restrictive hypothesis for phonological processes.

1 Introduction

A lot of recent work in computational phonology has taken the approach of representing phonological processes as *maps* or *functions* from input strings/underlying representations to output strings/surface representations (Chandlee and Heinz, 2012; Heinz and Lai, 2013; Chandlee, 2014; Luo, 2017; Payne, 2017). The question of interest underlying such work is how computationally powerful or expressive the functions involved in phonology need to be. In particular, these authors have argued that various proper subsets of the regular relations (i.e., *subregular* functions) are sufficient to model the attested range of phonological phenomena. Special attention has been given to one such subclass, the *subsequential* functions, which are restrictive in being deterministic but still sufficiently expressive to capture the needed range of segmental phonological processes, both local and long-distance.

While this work has largely proceeded using the finite-state formalism, a related line of work has aimed to address similar questions using logical characterizations (Jardine, 2016; Chandlee and Jardine, 2019), one advantage of logic being that it offers a unified approach to modeling both linear (i.e., string-based) and non-linear (i.e., autosegmental representations, metrical structure) phenomena. However, a restrictive logic that is still sufficiently expressive to cover a wide range of phonological phenomena has yet to be established. Recent work has studied the relationships between classes of automata and logical transductions (Engelfriet and Hoogeboom, 2001; Filiot, 2015; Filiot and Reynier, 2016), but no work has studied a logical characterization of the subsequential functions.

This paper thus aims to fill this gap by proposing a type of logic that approximates the subsequential class of functions in an important way. This logic is quantifier-free least fixed point (QFLFP), which is first order logic *without* quantifiers but *with* a (monadic) least fixed point operator. We will define this type of logic and demonstrate how it can be used to model a range of phonological processes in a recursive and output-oriented way. We will also show that the functions can be defined with this type of logic are in fact a proper subset of the subsequential functions. This is interesting both because it shows how logical transductions can be related to subsequential functions, and also because it closely fits the attested typology of phonological processes.

The paper is organized as follows. In §2 we provide the needed preliminaries, and in §3 we give an overview of the hierarchy of function classes in the finite-state formalism. In §4 we provide relevant background on logical characterizations of functions, quantifier-free logic, and least fixed point operators, before defining QFLFP functions. In §5 we demonstrate the application of QFLFP functions to phonology by giving example analyses of phonological processes. In §6 we establish certain properties of QFLFP, and in §7 we discuss the implications of our findings and highlight a few areas for future work. §8 concludes.

2 Preliminaries

Given a finite alphabet Σ, with Σ^* we designate the set of all possible finite strings of symbols from Σ and with $\Sigma^{\leq n}$ we designate the set of all possible finite strings of length $\leq n$. The length of a string w is $|w|$. The unique empty string is λ, so $|\lambda| = 0$.

Let $\mathcal{P}(X)$ denote the powerset of a set X.

For a set of strings L, $\mathrm{pref}(L) = \{u \mid \exists v \text{ such that } uv \in L\}$, $\mathrm{suff}(L) = \{v \mid \exists u \text{ such that } uv \in L\}$, and for a string w, $\mathrm{pref}(w) = \{u \mid w = uv \text{ for some string } v\}$,

$$\mathrm{pref}_n(w) = \begin{array}{ll} w & \text{if } |w| \leq n; \text{ otherwise} \\ w_1, & \text{where } w = w_1 w_2, |w_1| = n, \end{array}$$

$$\mathrm{suff}_n(w) = \begin{array}{ll} w & \text{if } |w| \leq n; \text{ otherwise} \\ w_2, & \text{where } w = w_1 w_2, |w_2| = n. \end{array}$$

For a string $w = \sigma_1 ... \sigma_i ... \sigma_n$, let $w(i)$ denote σ_i.

For a set of strings $L \subseteq \Sigma^*$ and two strings $w, v \in \Sigma^*$, we write $w \equiv_L v$ iff $\forall z \in \Sigma^*$, $wz \in L \leftrightarrow vz \in L$. This is an equivalence relation on Σ^* inducing a partition P_L on Σ^*. A set L is *regular* if and only if P_L is finite. Note that if $w, v \in A$ for some $A \in P_L$, then for any $z \in \Sigma^*$ then $wz \in A' \leftrightarrow vz \in A'$ for some other $A' \in P$.

We can define a similar notion for functions on Σ^*. For a set of strings L, the *longest common prefix* is the longest shared prefix of all the strings in L. Formally,

$$\mathrm{lcp}(L) = \begin{array}{l} u \in \bigcap_{w \in L} \mathrm{pref}(w), |u| \geq |u'| \\ \text{for all } u' \in \bigcap_{w \in L} \mathrm{pref}(w). \end{array}$$

For a function $f : \Sigma^* \to \Gamma^*$, and a string $w \in \Sigma^*$, the *tails* of w with respect to f are defined as follows:

$$\mathrm{tails}_f(w) = \{(z, z') \mid f(wz) = uz', \\ u = \mathrm{lcp}(f(w\Sigma^*))\}.$$

Two strings $w, v \in \Sigma^*$ are *tail-equivalent*, written $w \equiv_f v$ iff $\mathrm{tails}_f(w) = \mathrm{tails}_f(v)$. Note that $\equiv_f$ is an equivalence relation on Σ^*; let P_f be the partition it induces on Σ^*. A function f is *subsequential* (SUBSEQ) iff P_f is finite (Oncina et al., 1993).

We will make use of the following operation that takes the pairwise intersection of the blocks of two partitions P_1 and P_2 on some set X.

$$P_1 \otimes P_2 \overset{\mathrm{def}}{=} \{A \cap B \neq \varnothing \mid A \in P_1, B \in P_2\}$$

It is elementary to show that $P_1 \otimes P_2$ is also a partition on X, and that the operation is associative and commutative.

3 Transducers and function classes

It will be helpful to first define some important function classes and illustrate them with automata. The SUBSEQ class is exactly described by *subsequetial finite-state transducers* (SFSTs), or determinstic finite state machines that output a string for each input symbol and upon ending on a state (Schützenberger, 1977; Mohri, 1997). Note that this makes SUBSEQ a strict sublcass of the *regular* (or *rational*) functions, or exactly those describable with an FST (deterministic or otherwise). Formally, a SFST is a tuple $\langle Q, q_0, Q_f, \Sigma, \Gamma, \delta, \omega \rangle$ where Q is the set of states, $q_0 \in Q$ is the single initial state, $Q_f \subseteq Q$ is the set of final states, Σ and Γ are the input and output alphabets, respectively, $\delta : (Q \times \Sigma) \to (\Gamma^* \times Q)$ is a *transition function*, and $\omega : Q_f \to \Gamma^*$ is the *output function*. Note that because δ takes pairs in $(Q \times \Sigma)$ (that is, it does not have transitions on an input λ), and because it is a function, the machine is deterministic. We define the transitive closure δ^* of δ recursively in the usual way; i.e. $\delta^*(q, \lambda) = (\lambda, q)$ and if $\delta^*(q, w) = (v_1, q_1)$ and $\delta(q_1, \sigma) = (v_2, q_2)$ then $\delta^*(q, w\sigma) = (v_1 v_2, q_2)$.

A transducer $T = \langle Q, q_0, Q_f, \Sigma, \Gamma, \delta, \omega \rangle$ describes a function $f_T : \Sigma^* \to \Gamma^*$ defined as $f_T(w) = uv$ where $\delta^*(q_0, w) = (u, q_f)$ for some $q_f \in Q_f$, and $v = \omega(q_f)$. As an example, the SFST in Fig 1 models the function described by the rewrite rule in (1) ('change an a that follows a b to b'). (In all SFSTs in this paper, the initial state is indicated with a small unlabeled incoming arrow, all states are assumed to be final, and the output function is represented in the state label with the string to the right of the colon.)

(1) $a \to b \,/\, b \underline{}$ (simultaneous)

As stated above, SUBSEQ is exactly the class of functions with a finite set of tail-equivalence classes. In the minimal SFST for a subsequential function, each state corresponds to a tail-equivalence class.

Restrictions on the nature of these tail-equivalence classes offer *subclasses* of SUBSEQ that have been argued to be relevant to phonology. One such class is the *input strictly local* (ISL) functions, a subsequential

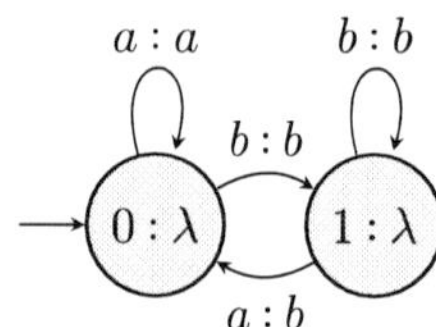

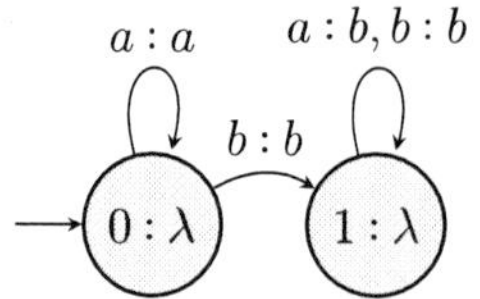

Figure 1: A SFST implementing the rule a → b / b __, applied simultaneously, for $\Sigma = \Gamma = \{a, b\}$. This function is ISL$_2$.

Figure 2: A SFST implementing the rule a → b / b __, applied iteratively, for $\Sigma = \Gamma = \{a, b\}$. This function is OSL$_2$.

class defined by Chandlee (2014), which Chandlee and Heinz (2018) argue can model any local phonological rule that applies simultaneously. Formally, the ISL class is defined as below.

Definition 1 (ISL; Chandlee 2014) *A function f is ISL$_k$ iff for all strings w, v,* $\mathtt{suff}_{k-1}(w) = \mathtt{suff}_{k-1}(v)$ *implies* $\mathtt{tails}_f(w) = \mathtt{tails}_f(v)$.

Intuitively, the ISL class is exactly those functions for which, for any symbol in the string, its output is entirely decided by the preceding $k - 1$ symbols in the input. This means that any ISL$_k$ function can be described by a SFST whose states correspond to the $k - 1$ suffixes of Σ^*. Fig. 1 is exactly one such SFST: its states represent the previous 1 symbol in the input—state 0 represents a preceding a in the input, and state 1 a preceding b. The function is thus ISL$_2$.

Another notion of local string functions is those for which the output of a string is based on the preceding $k - 1$ symbols in the *output*; this is the *output strictly local* (OSL) class. A definition for this class is given below.

Definition 2 (OSL; Chandlee 2014)
A function f is OSL$_k$ iff for all strings w, v, $\mathtt{suff}_{k-1}(f(w)) = \mathtt{suff}_{k-1}(f(v))$ *implies* $\mathtt{tails}_f(w) = \mathtt{tails}_f(v)$.

(Note: this is an incomplete definition, but for the purposes of this paper it is sufficient. For the complete definition see Chandlee et al. 2015.) The definition for OSL is parallel to that for ISL, except it refers to the output of the function. An example is given in Fig. 2 for the rule in (2), applied iteratively.

(2)   a → b / b __ (iterative)

By "applied iteratively," we mean that an a becomes a b after a b which may have been present in the input (and so remained a b in the output) or

may have been an a in the input (and so was itself turned into a b in the output). This is OSL$_2$, because (either way) whether or not an input a is output as b depends on the immediately preceding *output* symbol.

The way in which ISL and OSL correspond to the difference between simultaneous and iterative application of rules can be seen more clearly when one compares the output strings for these two FSTs for the input *baabaa*. The FST in Fig. 1 outputs *bbabba* for this input string, such that all a's that follow b's *in the input* are changed to b. In contrast, the FST in Fig. 2 outputs *bbbbbb*, such that all a's are changed to b. The structural difference in the FSTs that is responsible for this difference is that in Fig. 1 all transitions with the same input symbol lead to the same state whereas in Fig. 2 all transitions with the same output string lead to the same state.[1]

The ISL and OSL functions are strict subclasses of SUBSEQ. That ISL $\subsetneq$ SUBSEQ is witnessed by Fig. 3. The SFST in this figure represents a function that changes a to b if it follows a b, where any number of c's can intervene. The strings c^k and bc^k clearly have the same suffix of length $k - 1$ (for any k), but (contra Definition 1) they have different tails, as (a, a) is in the tails of c^k while (a, b) is in the tails of bc^k.

That OSL $\subsetneq$ SUBSEQ is also witnessed by Fig. 3. The strings $f(c^k) = c^k$ and $f(bc^k) = bc^k$ again have the same suffix of length $k - 1$ (for any k), but (contra Definition 2) they have different tails, as (a, a) is in the tails of c^k while (a, b) is in the tails of bc^k.

The ISL and OSL classes are thus restrictive classes of functions that have been posited as strong hypotheses for the computational complexity of phonological processes with local triggers (Chandlee, 2014; Chandlee and Heinz, 2018). Examples of their connection to phonological pro-

---

[1]For more on how FSTs model modes of rule application see Kaplan and Kay (1994) and Hulden (2009).

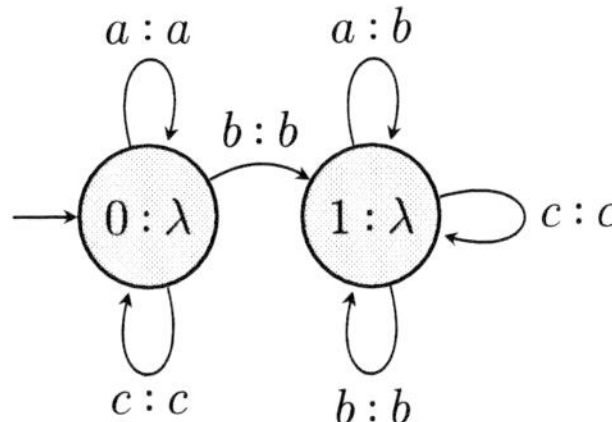

Figure 3: A SFST changing any a following a b in a word, regardless of any number of intervening c's, to a b. This function is neither ISL or OSL.

cesses are given below. The purpose of this paper is to investigate, through logical transductions, a class that generalizes the intuitions behind these classes, and also captures non-ISL and non-OSL functions, like Fig. 3, that are relevant to phonology.

4 Logical transductions

We consider finite strings models over the signature $\mathcal{I}$ of strings in Σ with predecessor function p, delimited by boundaries $\rtimes$ and $\ltimes$, and assume that for a string of length n (including the boundaries) the domain of its model is $D = \{1, ..., n\}$.

$$\mathcal{I} = \{p, \rtimes, \ltimes, P_{\sigma \in \Sigma}\}$$

Figure 4 gives a model for a string over the alphabet $\Sigma = \{a, b\}$ with $D = \{1, 2, 3, 4, 5, 6, 7\}$. For each $\sigma \in \Sigma \cup \{\rtimes, \ltimes\}$, $P_\sigma \subseteq D$ includes those members of D that are labeled with σ. For example, in the figure $P_b = \{3, 4, 6\}$.

$\rtimes_1$	a_2	b_3	b_4	a_5	b_6	$\ltimes_7$

Figure 4: Model for the string *abbab*

The signature also includes a predecessor function p that gives the immediately proceeding position for each position—e.g., $p(3) = 2, p(2) = 1$, etc.

We define a monadic second order (MSO) logic $\mathcal{L}_\mathcal{I}$ over $\mathcal{I}$ in the usual way; that is, $\mathcal{L}_\mathcal{I}$ is a predicate logic whose atomic formulae are of the form $\sigma(t)$, where $\sigma(t)$ is true when the *term* t is interpreted as a member of P_σ. More formally, a term is either some member x of a countably infinite set of first-order variables (which range over the domains of models) or the application $p(t)$ of the predecessor function to a term t. MSO is then a logic defined recursively in which $\sigma(t)$ for some t

and $\sigma \in \Sigma$ is a *well-formed formula* (WFF); $X(t)$ for some term t and some variable X drawn from a countably infinite set of set variables is a WFF; for WFFs φ and ψ, $\neg \varphi$ and $\varphi \vee \psi$ are WFFs; and for WFFs $\varphi(x)$ and $\psi(X)$ with free first-order and set variables x and X, respectively, $(\exists x)[\varphi(x)]$ and $(\exists X)[\psi(X)]$ are WFFs. As shorthand, for $k \geq 0$, $p^k(x)$ denotes x when $k = 0$, $p(p^{k-1}(x))$ otherwise. The semantics of $\mathcal{L}_\mathcal{I}$ are defined as usual over string models with the first-order variables mapped to positions in a string and second-order variables mapped to sets of positions in the string. In particular we write $\mathcal{S} \models \varphi(x)[x \mapsto i]$ for a string model $\mathcal{S}$ that satisfies $\varphi(x)$ when x is mapped to $i \in D$.

For example, $b(p(x))$ is a WFF in $\mathcal{L}_\mathcal{I}$, as is $b(x) \wedge b(p(x))$. The string in Fig. 4 satisfies $b(p(x))$ when x is mapped to positions 4, 5, or 7 (that is, the positions whose predecessor is labeled b). The string in Fig. 4 satisfies $b(x) \wedge b(p(x))$ only when x is mapped to 4, as that is the only position that is both a b and is immediately preceded by a b.

We can define functions logically through a *logical interpretation* of an output signature $\mathcal{O}$ in the logic $\mathcal{L}_\mathcal{I}$ of the input signature. Specifically, a MSO transduction T to the output signature $\mathcal{O}$ over strings in Γ

$$\mathcal{O} = \{<', P_{\gamma \in \Gamma}\}$$

is a finite ordered *copy set* $C = \{1, ..., n\}$ and a series of formulae $\gamma^c(x)$ in $\mathcal{L}_\mathcal{I}$ with exactly one free variable x and for each $\gamma \in \Gamma$ and $c \in C$. The copy set allows us to create up to n output copies of each input element.[2] See below for an example demonstrating the use of multiple copies of an input element.

We define the semantics of T following Engelfriet and Hoogeboom (2001). For an input string model $\mathcal{S}$ over the signature $\mathcal{I}$ with domain D, its output $T(\mathcal{S})$ is a string $\mathcal{S}'$ over the signature $\mathcal{O}$ with domain D' and unary relations $P_{\gamma \in \Gamma}$ built in the following way. For each $d \in D$, there is a $d^c \in D'$ that belongs to P_γ if and only if $\mathcal{S} \models \gamma^c(x)[x \mapsto d]$ for exactly one $\gamma \in \Gamma$ and exactly one $c \in C$. If no such γ or c exists, then no output position is constructed for d. We assume

---

[2] We can also consider a closed domain formula φ_{dom} which specifies the domain of the function; as long as this formula is in MSO or lower, this does not change any of the below results, so we do not discuss it in detail here. For more see Engelfriet and Hoogeboom (2001); Filiot (2015).

that the transduction is *order-preserving*; that is, the output order $<'$ in $\mathcal{O}$ is such that for any $d \in D$ and $i < j \in C$, $d^i <' d^j$, and for any distinct $d, e \in D$ and $i, j \in C$, for the copies d^i and e^j in D', $d^i <' e^j$ if and only if $d < e$.

Thus, a MSO transduction T thus defines a function $f(T)$ from strings in $\rtimes \Sigma^* \ltimes$ to strings in Γ^*.

As an example, we logically define the function for 'change a b following another b to a c' rule given in (3), assuming $\Sigma = \{a, b\}$ and $\Gamma = \{a, b, c\}$.

(3) $b \to c \,/\, b\_\_$

Since this rule makes no mention of a's, all positions in the input model that are labeled a can likewise be labeled a in the output model. The formula in (4-a) achieves this. As this is a 'substitution' rule that doesn't extend the length of the string, only a single copy of each input position is needed. Therefore we simply mark the output relations with a prime γ' instead of a number.

(4)   a. $a'(x) \overset{\text{def}}{=} a(x)$

       b. $b'(x) \overset{\text{def}}{=} b(x) \land \neg b(p(x))$

       c. $c'(x) \overset{\text{def}}{=} b(x) \land b(p(x))$

Likewise, the formula in (4-b) declares which positions in the output model are labeled b: namely, those positions that are labeled b in the input model but whose predecessors are *not* also labeled b (so excluding those b's that are subject to the rule in (3)). Lastly, the formula in (4-c) declares which positions are labeled c in the output model: those positions that are subject to the rule in (3). The collective result is the output model shown in Figure 5.

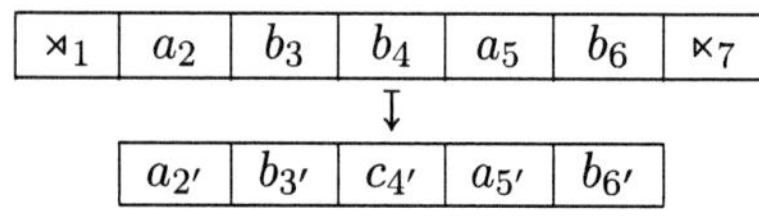

Figure 5: Transduction of *abbab* following (4).

The following example demonstrates the use of a copy set with a size larger than 1. With the following formulas and a copy set $C = \{1, 2\}$ we define a function that inserts a c after every input b.

(5)   a. $a^1(x) \overset{\text{def}}{=} a(x)$

       b. $a^2(x) \overset{\text{def}}{=} \texttt{False}$

       c. $b^1(x) \overset{\text{def}}{=} b(x)$

       d. $b^2(x) \overset{\text{def}}{=} \texttt{False}$

       e. $c^1(x) \overset{\text{def}}{=} \texttt{False}$

       f. $c^2(x) \overset{\text{def}}{=} b(x)$

The formulas (5-a) and (5-c) create a first copy labeled a and b in the output, respectively, for each a and b in the input. The formula (5-f), then, creates a second copy labeled c for each b in the input. As the other formula are set to `False`, no other copies are produced in the output. Finally, following the precedence relation defined above, each 1 copy b precedes the 2 copy c. Thus, *abbab* is output as *abcbcabc*.

Fig. 6 illustrates the transduction defined by these formulas with the input string *abbab*.

⋊1	a2	b3	b4	a5	b6	⋉7

↓

a2¹	b3¹	b4¹	a5¹	b6¹
	c3²	c4²		c6²

Figure 6: Transduction of *abbab* into *abcbcabc* following (5)

We fix the precedence relations in the output due to the following equivalence.

Theorem 1 (Filiot 2015) *A function f is regular iff there is an order-preserving monadic second order transduction T such that $f = f(T)$.*

Theorem 1 thus guarantees that any logic and signature we use that is less than or equal to monadic second order logic in expressivity will give us (sub-)regular functions.

4.1 Quantifier-free logic

Note that none of the formulas used in the previous example included quantifiers. Let *quantifier-free* (QF) denote the restriction on $\mathcal{L}_\mathcal{I}$ to formula with no quantifiers (and thus no set variables).

Chandlee and Lindell (in prep.) relate QF to the ISL class, showing that any ISL function with bounded deletion is QF-definable. However, their definitions do not consider functions with the output order as defined above. We show in §6.2 that, given this order, QF = ISL exactly.

An example of a phonological rule that cannot be modeled with this restricted logic is unbounded iterative spreading of a feature, such as the spreading of nasality from a nasal consonant to a following sequence of vowels: $nVVV \mapsto n\tilde{V}\tilde{V}\tilde{V}$. The formula in (6) would achieve the nasalization of the first vowel following the input nasal.

$$(6) \qquad \tilde{V}'(x) \stackrel{\text{def}}{=} n(p(x))$$

The problem with modeling this process using only QF formula arises when trying to account for the nasalization of the second and third vowels. Of course we can add to (6) $n(p^2(x)) \vee n(p^3(x))$, which would accommodate the present example. But then what about a form like /nVVVV/, for which the nasalization of the final vowel would require a formula to identify the nasality of the predecessor of its predecessor of its predecessor of its predecessor (i.e., $p^4(x)$)? The issue now should be clear: for an unbounded pattern such as this one, the input nasal that triggers the nasalization of a given vowel may be an unbounded number of segments away. Without knowing in advance how large of a bound that is, a formula in terms of the predecessor function cannot be defined. Instead, what is needed is a quantifier: nasalize a vowel if there *exists* a nasal consonant at a prior position.

The traditional analysis of a feature spreading process like this one is that after the first nasalization of the vowel immediately following the nasal consonant, the actual trigger of the next nasalization is the most 'recently' nasalized vowel. But capturing that intuition requires specifying something about the labels in the *output* model, not the input model, which falls beyond the capabilities of QF. In that sense the limitation here is not the need for the quantifier, but the restriction on the formula to referring to the input model only.

In the next section we will show how to model such output-oriented processes using QF formula with the addition of the *least fixed point* (`lfp`) operator.

4.2 Least fixed point quantifier-free logic

Least fixed point logic allows us to add inductive definitions to our logics. The following is based on Libkin (2004, Ch. 10), but simplified for unary predicates.

For a set U an *operator* on U is a function $F : \mathcal{P}(U) \to \mathcal{P}(U)$. A set $X \subseteq U$ is a *fixed point* if $F(X) = X$. A set X is the *least fixed point* $\text{lfp}(F)$ of F iff it is a fixed point of F and for every other fixed point Y of F, $X \subseteq Y$. We write $\text{lfp}(F)$ for the least fixed point of F.

An operator F is *monotone* if $X \subseteq Y$ implies $F(X) \subseteq F(Y)$. For every monotone operator F,

$\text{lfp}(F) = \bigcup_i X^i$, where each X^i is from the sequence in (7).

$$(7) \qquad X^0 = \varnothing, X^{i+1} = F(X^i)$$

That is, $\text{lfp}(F)$ is the set that is converged to by recursive applications of F. Given models with finite domains, there is always such a (finite) set for a monotone operator on the domain.

For a signature $\mathcal{S}$ and any model $\mathcal{M}$ in the signature whose universe is M, we extend our logic with an additional set predicate A not in $\mathcal{S}$. A formula $\varphi(A, x)$ with a single free variable x and free set variable A induces an operator F_φ on M as follows.

$$F_\varphi(X) = \{m \mid \mathcal{M} \models \varphi(A, x)[A \mapsto X, x \mapsto m]\}$$

We will work through an example to illustrate, using the string model in Figure 7. Let $\varphi(A, x)$ be defined as in (8).

$\rtimes_1$	a_2	b_3	a_4	a_5	a_6	c_7	a_8	$\ltimes_9$

Figure 7: Model for the string *abaaaca*

$$(8) \qquad \varphi(A, x) = a(x) \wedge (b(p(x)) \vee A(p(x)))$$

First let $X^0 = \varnothing$. We then have

$$F_\varphi(\varnothing) = \{m \mid D \models \varphi(A, x)[A \mapsto \varnothing, x \mapsto m]\}.$$

No position $m \in D$ satisfies $A(p(x))$ since $A = \varnothing$, but position 4 satisfies both $a(x)$ and $b(p(x))$, and so $F_\varphi(\varnothing) = \{4\}$.

Now $X = \{4\}$, so $F_\varphi(\{4\}) = \{m \mid D \models \varphi(A, x)[A \mapsto \{4\}, x \mapsto m]\}$. Now position 5 satisfies the formula (because it is labeled a and its predecessor is in A), so $F_\varphi(\{4\}) = \{4, 5\}$.

Now with $X = \{4, 5\}$, $F_\varphi(\{4, 5\}) = \{4, 5, 6\}$, and then $F_\varphi(\{4, 5, 6\}) = \{4, 5, 6\}$, so $\{4, 5, 6\}$ is $\text{lfp}(F_\varphi)$, as no new elements are added.

Whether or not F_φ is monotone (and thus, whether $\text{lfp}(F_\varphi)$ can be determined via the sequence in (7)) is undecidable for an arbitrary FO formula φ. So the least fixed point extension of FO is defined with the following restriction on the set variable A:

$(9) \qquad A$ is *positive* in φ; that is, it is under the scope of an even number of negations.

We then extend the usual definition of QF with the following syntax:

Definition 3 (QFLFP **syntax**) *Any formula in* QF *is in* QFLFP.

For a formula $\varphi(A, y)$ in QF *extended with the predicate $A(y)$ satisfying (9),* $[\texttt{lfp}\ \varphi(A, y)](x)$ *is a formula in* QFLFP.

A formula $[\texttt{lfp}\ \varphi(A, y)](x)$ is then true when x is interpreted as an element in the least-fixed point of the operator induced by $\varphi(A, y)$.

Definition 4 (QFLFP **semantics**) *For any formula in* QFLFP *that is also in* QF, *satisfaction is the same as in* QF.

For a formula $[\texttt{lfp}\ \varphi(A, y)](x)$ *in* QFLFP,

$$\mathcal{M} \vDash [\texttt{lfp}\ \varphi(A, y)](x)[x \mapsto m] \text{ iff } m \in \texttt{lfp}(F_\varphi)$$

In the next section we will demonstrate such transductions with phonological examples.

5 QFLFP functions in phonology

First we return to the example of unbounded iterative spreading, discussed in the previous section as a case that can't be modeled as a QF transduction. For simplicity we will assume the alphabet $\Sigma = \{a, b\}$ and use the model in Figure 8.

⋊$_1$	b_2	a_3	a_4	a_5	⋉$_6$

Figure 8: Model for the string *baaa*

The rule is the same as in (2), repeated below in (10).

(10) $a \rightarrow b\ /\ b\_\_$ (iterative)

The expected output string is then *bbbb*. The formula in (11) uses a $\texttt{lfp}$ operator to declare which output positions should be labeled b.

(11) $b'(x) \overset{\text{def}}{=} [\texttt{lfp}\ b(y) \vee A(p(y))](x) \wedge \neg ⋉ (x)$

At first only position 2 satisfies $b(y) \vee A(p(y))$, since initially $A = \varnothing$ and so no positions satisfy $A(p(y))$. Now with $A = \{2\}$, position 3 satisfies $b(x) \vee A(p(y))$, since its predecessor is in A. Next 4 can be added, and then 5, and finally 6. After that no new positions will ever be added, and so $\{2, 3, 4, 5, 6\}$ is the $\texttt{lfp}$. The copies of theses positions are all labeled b in the output—with the exception of 6 due to $\neg ⋉ (x)$. Thus the map *baaa* $\mapsto$ *bbbb* is obtained.

We also consider a case of unbounded spreading with blocking, in which the feature spreads unboundedly but only up to a particular type of segment, a blocker. This type of process is attested in Johore Malay (Onn, 1980), where nasality spreads from a nasal consonant through a following span of vowels and glides but stops when it reaches an obstruent:

(12)   Johore Malay (Onn, 1980)
/pəŋawasan/ ↦ [pəŋā̃wã̃san], 'supervision'

We will extend our analysis of unbounded spreading to this case as well, using the alphabet $\Sigma = \{a, b, c\}$, where c is a blocker.

⋊$_1$	b_2	a_3	a_4	c_5	a_6	b_7	a_8	⋉$_9$

Figure 9: Model for the string *baacaba*

As (13) shows, the change is minor: the conditions under which a position is added to the set are made more stringent by including $\neg c(x)$.

(13) $b'(x) \overset{\text{def}}{=} [\texttt{lfp}(b(y) \vee (A(p(y)) \wedge \neg c(y)))](x) \wedge \neg ⋉ (x)$

The effect of this added restriction is that the set will be built as follows:

$\varnothing$
$\{2, 7\}$
$\{2, 3, 7, 8\}$
$\{2, 3, 4, 7, 8, 9\}$

This last set is the $\texttt{lfp}$, which picks out those positions that will be labeled b in the output (again except for 9), giving the string *bbbcabb*. The a that follows the blocker c will not receive the spreading feature because it is never added to the set, because its predecessor is prevented from joining the set by being labeled c.

Lastly, we consider a case of unbounded agreement, where a segment takes on a feature from a triggering segment but (unlike in spreading) the intervening segments are unaffected. An example is in Kikongo, in which a liquid becomes nasal (underlined below) following a nasal somewhere in a root (Ao, 1991; Odden, 1994).

(14)   Kikongo (Ao, 1991)

a. /ku-toot-ila/ $\quad \mapsto$ [ku-toot-ila]
   'to harvest for'

b. /ku-dumuk-ila/ $\mapsto$ [ku-dumuk-i<u>na</u>
   'to jump for'

c. /ku-dumuk-is-ila/$\mapsto$[ku-dumuk-is-i<u>na</u>]
   'to make jump for'

We can model this process schematically with the function from Fig. 3, in which any a following a b is output as a b, no matter how many c's intervene. We will use the string model in Figure 10, where b is the trigger, a is the target, and c intervenes (without blocking).

$$\boxed{\rtimes_1} \; \boxed{c_2} \; \boxed{b_3} \; \boxed{c_4} \; \boxed{c_5} \; \boxed{c_6} \; \boxed{a_7} \; \boxed{\ltimes_8}$$

Figure 10: Model for the string $cbccca$

This time the positions labeled b in the output are identified by the formula in (15).

$$(15) \qquad b'(x) \stackrel{\text{def}}{=} [\texttt{lfp}\ b(y) \vee A(p(y))](x) \wedge \neg c(x) \wedge \neg \ltimes (x)$$

The set is built as follows:

$\varnothing$
$\{3\}$
$\{3,4\}$
$\{3,4,5\}$
$\{3,4,5,6\}$
$\{3,4,5,6,7\}$
$\{3,4,5,6,7,8\}$

Note that all positions after the trigger are included in the set, but the formula for $b'(x)$ specifies that to be labeled b in the output a position has to both be in this set AND not be labeled c (or again $\ltimes$). So only the trigger and target are labeled b, giving the output string $cbcccb$.

6 Properties of QFLFP

6.1 Relation of QFLFP to other classes of functions

We now show some important properties of QFLFP. In particular, the structure of QFLFP concisely restricts its transductions to subsequential functions. Specifically, QFLFP transductions model the deterministic computation of the output string reading the input string left to right.

Remark 1 QFLFP $\subseteq$ MSO.

Proof: It is sufficient to show a translation into MSO for formulas of the form $[\texttt{lfp}\ \varphi(A,y)](x)$. We can replace any such formula for an equivalent MSO formula $(\exists X \forall y)[(\varphi(A/X,y) \rightarrow X(y)) \wedge X(x)]$, where $\varphi(A/X,y)$ is $\varphi(A,y)$ with each instance of $A(p^n(y))$ replaced with $X(p^n(y))$. $\qquad\square$

The following shows that satisfaction of a QFLFP formula is closed under suffixation; that is, if a position in a string satisfies a formula, it will satisfy that formula regardless of how the string is suffixed. This shows that QFLFP is strictly less powerful than full MSO transductions.

Lemma 1 *For any QFLFP formula* $[\texttt{lfp}\ \varphi(A,y)](x)$, *any* $w \in \texttt{pref}(\rtimes \Sigma^* \ltimes)$, *and any* $1 \leq n \leq |w|$,

$$w \vDash [\texttt{lfp}\ \varphi(A,y)](x)[x \mapsto n]$$
$$\text{implies}$$
$$wv \vDash [\texttt{lfp}\ \varphi(A,y)](x)[x \mapsto n]$$

for all $v \in \texttt{suff}(\rtimes \Sigma^* \ltimes)$.

Proof: Let $\texttt{lfp}_w(F_\varphi)$ be the least fixed point with respect to F_φ for the domain D_w of (the model of) w; likewise $\texttt{lfp}_{wv}(F_\varphi)$ for wv. We show that $\texttt{lfp}_w = (\texttt{lfp}_{wv} \cap D_w)$; this means that $n \in \texttt{lfp}_w$ if and only if $n \in \texttt{lfp}_{wv}$ for all $1 \leq n \leq |w|$.

Consider the series $X_w^0 = \varnothing$, $X_w^{i+1} = F_\varphi(X_w^i)$ from (7), relativized to the domain of w, and similarly $X_{wv}^0 = \varnothing$, $X_{wv}^{i+1} = F_\varphi(X_{wv}^i)$ for wv. By definition $X_w^0 = X_{wv}^0 = \varnothing$.

It is then the case that $X_w^1 = (X_{wv}^1 \cap D_w)$. For any $1 \leq j \leq |w|$, $w \vDash \varphi(A,y)[A \mapsto \varnothing, y \mapsto j]$ implies $wv \vDash \varphi(A,y)[A \mapsto \varnothing, y \mapsto j]$ because they are equivalent with respect to the base cases: for $\varphi(A,y) = a(p^m(y))$, $w \vDash a(p^m(y))[y \mapsto j]$ iff $v \vDash a(p^m(y))[y \mapsto j]$ because $\texttt{pref}_j(w) = \texttt{pref}_j(wv)$; $w,v \nvDash A(p^m(y))[A \mapsto \varnothing, y \mapsto j]$, because this always evaluates to false. So $j \in X_w^1$ iff $j \in X_{wv}^1$ for $1 \leq j \leq |w|$.

We then consider X_w^i and X_{wv}^i for an arbitrary $i \geq 2$. By hypothesis, assume that $X_w^i = (X_{wv}^i \cap D_{wv})$. Then for any $1 \leq j \leq |w|$, because $\texttt{pref}_j(w) = \texttt{pref}_j(wv)$, $w \vDash \varphi(A,y)[A \mapsto X_w^i, y \mapsto j]$ iff $v \vDash \varphi(A,y)[A \mapsto X_{wv}^i, y \mapsto j]$. So $j \in X_w^{i+1}$ iff $j \in X_{wv}^{i+1}$, and thus $X_w^{i+1} = (X_{wv}^{i+1} \cap D_{wv})$.

Because we have shown this also for $i = 1$, it must then hold for any arbitrary i. Thus $\texttt{lfp}_w = (\texttt{lfp}_{wv} \cap D_w)$. $\qquad\square$

Lemma 2 *For any* QFLFP *formula* $\varphi(x)$, *any* $w \in \texttt{pref}(\rtimes\Sigma^*\ltimes)$, *and any* $1 \le n \le |w|$,

$$w \vDash \varphi(x)[x \mapsto n] \text{ implies } wv \vDash \varphi(x)[x \mapsto n]$$

for all $v \in \texttt{suff}(\rtimes\Sigma^*\ltimes)$.

Proof: By recursion on the structure of $\varphi(x)$. If $\varphi(x) = a(p^m(x))$ for some $a \in \Sigma \cup \{\rtimes, \ltimes\}$, then $w(n) = wv(n) = a$. If $\varphi(x) = [\texttt{lfp }\psi(A, y)](x)$ and $w \vDash \varphi(x)[x \mapsto n]$ then $wv \vDash \varphi(x)[x \mapsto n]$ by Lemma 1. The cases for $\varphi(x) = \neg\psi(x)$ and $\varphi(x) = \psi_1(x) \vee \psi_2(x)$ then follow. $\square$

The above means that for a QFLFP transduction T the output of a prefix $w \in \Sigma^*$ in the input remains constant, regardless of how we extend w. We formalize this output with $\texttt{out}_T(w) = w_1' w_2' ... w_\ell'$, where $\ell = |w|$ and each $w_i' = \gamma_1 \gamma_2 ... \gamma_n$, such that $w \vDash \gamma_1^1(x)[x \mapsto i]$, $w \vDash \gamma_2^2(x)[x \mapsto i]$, ..., $w \vDash \gamma_n^n(x)[x \mapsto i]$ for $\gamma_1^1(x), \gamma_2^2(x), ..., \gamma_n^n(x) \in T$.

The following is thus important in establishing the tails of w.

Corollary 1 *For any string* $w \in \Sigma^*$, *a* QFLFP-*definable transduction* T *and its function* $f = f(T)$, $\texttt{out}(w)$ *is a prefix of* $\texttt{lcp}(f(w\Sigma^*))$.

Proof: Let $\texttt{out}_T(w) = w_1' w_2' ... w_\ell'$ as above. The corollary follows from Lemma 2, because at each position i in w, i will satisfy the same formulas no matter how w is extended, so w_i' will be output no matter how w is extended. Thus, at the least, $w_1' w_2' ... w_\ell'$ will be output. $\square$

The following establishes the set of 'reach strings' for $\varphi(x) \in T$; that is, strings w such that some $w\sigma$ will satisfy $\varphi(x)$. This simulates the set of strings that will reach a particular state in a FST.

Definition 5 *For* $\varphi(x) \in T$, *let*

$$L_\varphi \stackrel{def}{=} \left\{ w \mid w\sigma \vDash \varphi(x)[x \mapsto |w\sigma|] \text{ for some } \sigma \in \Sigma \right\}$$

The following follows from the fact that QFLFP $\subseteq$ MSO.

Remark 2 L_φ *is regular.*

For $\varphi(x) \in T$, let P_φ be the finite partition on Σ^* induced by the equivalence relation of L_φ. Then for $T = \{\varphi_1(x), ..., \varphi_n(x)\}$, let

$$P_T \stackrel{def}{=} P_{\varphi_1} \otimes P_{\varphi_2} \otimes ... \otimes P_{\varphi_n}$$

This is the partition of Σ^* into sets of strings belonging to the same equivalence class for every L_{φ_i}. As each L_{φ_i} is regular, it follows that P_T is finite.

Lemma 3 *For some* QFLFP $T = \{\varphi_1(x), ..., \varphi_n(x)\}$ *and* $f = f(T)$, *for any* $w, v \in \Sigma^*$ *and* $A \in P_T$,

$$w, v \in A \ \leftrightarrow \ w \equiv_f v$$

Proof: ($\rightarrow$) Recall (from §2) that for any $\varphi(x) \in T$ and $B \in P_\varphi$, $w, v \in B$ iff for all $z \in \Sigma^*$, $wz \in L_\varphi$ iff $vz \in L_\varphi$. Because P_T is the pairwise intersection of all such sets, for any $A \in P_T$, $w, v \in A$ if and only if for all $z \in \Sigma^*$ and for all $\varphi(x) \in T$, $wz \in L_\varphi \leftrightarrow vz \in L_\varphi$.

Thus, for $w, v \in A$ in P_T if and only if for all $z \in \Sigma^*$ and for all $\varphi(x) \in T$,

$$wz\sigma \vDash \varphi(x)[x \mapsto |wz\sigma|] \ \leftrightarrow$$
$$vz\sigma \vDash \varphi(x)[x \mapsto |vz\sigma|]$$

Consider then any $w, v \in A$ for some $A \in P_T$ and any $u = \sigma_1\sigma_2 ... \sigma_m$. It follows that for each σ_i $w\sigma_1 ... \sigma_i \vDash \varphi(x)[x \mapsto (|w| + i)]$ if and only if $v\sigma_1 ... \sigma_i \vDash \varphi(x)[x \mapsto (|v| + i)]$.

Then consider $u' = u_1' u_2' ... u_n'$ where each $u_i' = \gamma_1^1 \gamma_2^2 ... \gamma_k^k$ s.t. $w\sigma_1 ... \sigma_i \vDash \gamma_j^j(x)[x \mapsto (|w| + i)]$ for $\gamma_1^1(x), ..., \gamma_k^k(x) \in T$. From Lemma 2 we know

$$f(wu) = \texttt{out}_T(w)u' \text{ and } f(vu) = \texttt{out}_T(v)u'.$$

From Corollary 1 it is also the case that $\texttt{out}_T(w)$ is a prefix of $\texttt{lcp}(f(w\Sigma^*))$ and $\texttt{out}_T(v)$ is a prefix of $\texttt{lcp}(f(v\Sigma^*))$. Let u_p' be the portion of u' that is in $\texttt{lcp}(f(w\Sigma^*))$; that is, $u' = u_p' u_s'$ where $\texttt{lcp}(f(w\Sigma^*)) = \texttt{out}_T(w)u_p'$. Because the above establishes that w and v share the same output suffixes, it is also the case that $\texttt{lcp}(f(v\Sigma^*)) = \texttt{out}_T(v)u_p'$. Thus any (u, u_s') is in $\texttt{tails}_f(w)$ if and only if it is also in $\texttt{tails}_f(v)$.

Thus $w, v \in A$ implies that $\texttt{tails}_f(w) = \texttt{tails}_f(v)$.

($\leftarrow$) The reverse is clear by considering that $(u, u_1' u_2' ... u_n') \in \texttt{tails}_f(w)$ if and only if for each $u_i = \gamma_1^1 \gamma_2^2 ... \gamma_k^k$, $w\sigma_1 ... \sigma_i \vDash \gamma_j^j(x)[x \mapsto (|w| + i)]$ for each $\gamma_j^j(x)$. If this is also the case for v then it must be the case that $w, v \in A$ for the same $A \in P_T$. $\square$

Theorem 2 QFLFP $\subseteq$ SUBSEQ.

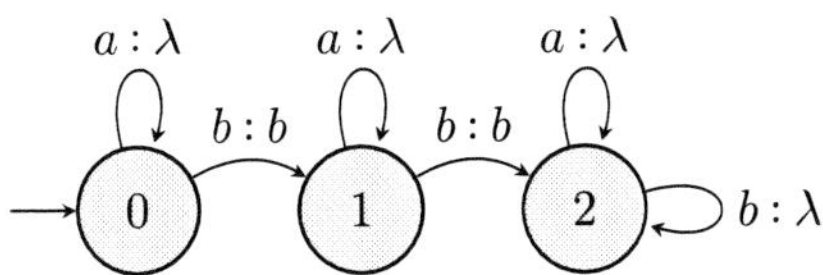

Figure 11: An OSL$_3$ function we conjecture not to be QFLFP.

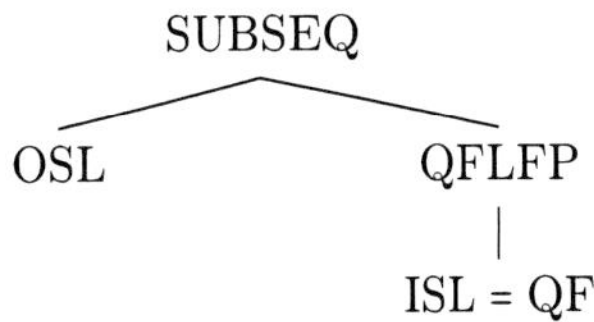

Figure 12: Hierarchy of the relevant function classes

Proof: From Lemma 3 and the fact that P_T is finite. $\qquad\square$

Lemma 4 QFLFP $\not\subseteq$ OSL

Proof: This is witnessed by the function of the SFST in Fig. 3 (which turns any a following a b into an a, regardless of any intervening c's), which was shown not to be OSL, but was shown in Sec. 5 to be QFLFP-definable. $\qquad\square$

Conjecture 1 OSL, SUBSEQ $\not\subseteq$ QFLFP

Consider the OSL function in Figure 11, which deletes any a and all except for the first two b's.[3] This is OSL$_3$, because whether or not we delete a b depends on whether or not we have output 2 b's previously. Now consider a formula $b'(x)$ that is true for exactly the first two b's. For the first, we can write $b(x) \wedge \neg[\texttt{lfp}(b(y) \vee A(p(y)))](x)$, the latter $\texttt{lfp}$ disjunct identifying any elements following a b. To identify exactly the second b, we would have to include reference to the first b in another $\texttt{lfp}$ predicate, thus embedding one $\texttt{lfp}$ statement in another. We conjecture that this embedding of one $\texttt{lfp}$ formula in another is necessary; that is, there is no QF formula $\varphi(A, y)$ such that $[\texttt{lfp}\,\varphi(A, y)](x)$ can identify the second b.

6.2 Equivalence of QF and ISL

Finally, we show that the QF fragment of QFLFP describes exactly the ISL functions. We first show that QF formula can only distinguish positions in a string based on the previous $k - 1$ symbols for some k.

Lemma 5 *Let T be a QF transduction, and let k be the value for which $p^{k-1}(x)$ appears in some $\gamma^c(x)$ in T and for any other $p^j(x)$ that appears in some $\gamma^c(x)$ in T, $j < k$. Consider two strings $w, v \in \texttt{pref}(\rtimes\Sigma^*\ltimes)$ such that $\texttt{suff}_{k-1}(w) = \texttt{suff}_{k-1}(v)$; let ℓ_w denote $|w|$ and likewise ℓ_v for*

$|v|$. *For any $\gamma^c(x)$ in T,*

$$w \vDash \gamma^c(x)[x \mapsto \ell_w]$$
$$\text{if and only if}$$
$$v \vDash \gamma^c(x)[x \mapsto \ell_v]$$

Proof: Let $u = \texttt{suff}_{k-1}(w) = \texttt{suff}_{k-1}(v)$ and ℓ_u denote $|u|$. If $\gamma^c(x) = a(p^j(x))$ for some $a \in \Sigma \cup \{\rtimes, \ltimes\}$, then $w \vDash \gamma^c(x)[x \mapsto \ell_w]$ iff $w(\ell_w - j) = a$. Since $j < k$, this implies that $u(\ell_u - j) = a$ and thus also that $v \vDash \gamma^c(x)[x \mapsto \ell_v]$. Clearly the reverse implication holds as well. Since w and v are equivalent with respect to satisfaction of the atomic formulae for QF, this then extends to the general case for $\varphi(x)$ based on induction on the structure of QF. $\square$

Theorem 3 QF = ISL

Proof: ($\rightarrow$) From Lemma 3 we know that w and v have the same set of tails; thus Lemma 5 shows that $\texttt{suff}_{k-1}(w) = \texttt{suff}_{k-1}(v)$ implies $w \equiv_f v$. Thus QF $\subseteq$ ISL.

($\leftarrow$) For ISL $\subseteq$ QF, we construct a QF transduction for the canonical transducer for an ISL function. As stated in §3, each state in an ISL$_k$ SFST represents a $k - 1$ suffix. So we can construct an equivalent QF transduction T as follows. For a string $w = \sigma_1\sigma_2...\sigma_k$ we can write $\varphi_w(x) = \sigma_1(p^{k-1}(x)) \wedge \sigma_2(p^{k-2}(x)) \wedge ... \wedge \sigma_{k-1}(p(x)) \wedge \sigma_k(x)$. For each symbol $\gamma \in \Gamma$, we set $\gamma^i(x) = \varphi_{w_1}(x) \vee \varphi_{w_2}(x) \vee ... \vee \varphi_{w_n}(x)$, where $w_1, w_2, ..., w_n$ is the exhaustive set of strings $w_j = q_j\sigma_j$ for which a transition $\delta(q_j, \sigma_j) = (v_j, r_j)$, exists where $\gamma = v_j(i)$. So for any position in the input string that exercises a transition $\delta(q, \sigma) = (v, r)$ in the ISL$_k$ machine, it will also satisfy a sequence $\gamma_1^1(x), \gamma_2^2(x), ..., \gamma_m^m(x)$ such that $\gamma_1\gamma_2...\gamma_m = v$. $\square$

Fig. 12 summarizes the results discussed so far.

---

[3] We thank Shiori Ikawa for this example.

6.3 Left- and right-subsequential functions

We have so far abstracted away from the fact that SUBSEQ can instead be characterized as the *left*-subsequential functions, with the *right*-subsequential functions being the reversal of some left-subsequential function (Mohri, 1997). Formally, a function f is right-subsequential iff $f = \{(w^r, v^r)|f'(w) = v\}$, where w^r is the reverse of string w, for some left-subsequential function f'. Equivalently, the right-subsequential functions are those that can be described by a SFST reading and writing strings right-to-left.

In terms of QFLFP, when the input signature $\mathcal{I}$ contains the predecessor function p we obtain left-subsequential functions. As is perhaps clear from the above discussion, if we instead use an identical signature with a successor function s, we obtain the right-subsequential functions.

Crucially, inclusion of QFLFP in one of the subsequential classes relies on using *either s or p*, but not *both*. Consider the following transduction defined with a signature including both s and p.[4]

$$(16) \quad \text{a. } a'(x) \stackrel{\text{def}}{=} a(x) \wedge \big(\rtimes (p(x)) \rightarrow$$
$$[\texttt{lfp}\ b(y) \vee A(s(y))](x)\big)$$
$$\text{b. } b'(x) \stackrel{\text{def}}{=} b(x) \wedge$$
$$[\texttt{lfp}\ b(y) \vee A(p(y))](x)$$
$$\text{c. } c'(x) \stackrel{\text{def}}{=} \big(a(x) \wedge \rtimes(p(x)) \wedge$$
$$\neg[\texttt{lfp}\ b(y) \vee A(s(y))](x)\big)$$
$$\vee$$
$$\big(b(x) \wedge$$
$$\neg[\texttt{lfp}\ b(y) \vee A(p(y))](x)\big)$$

This transduction takes strings of a's and b's and outputs as a c 1) the first a if and only if there are no b's in a string; or 2) the first b in the string. The definition of $c'(x)$ in (16-c) outputs a c for an input initial a that is not followed by a b (the first disjunct; $a(x) \wedge \rtimes(p(x)) \wedge \neg[\texttt{lfp}\ b(y) \vee A(s(y))](x))$ or an input b that is not preceded by another b (the second disjunct; $b(x) \wedge \neg[\texttt{lfp}\ b(y) \vee A(p(y))](x))$. The definitions in (16-a) and (16-b) output a's and b's, respectively, for any a and b that does not meet the conditions in (16-c) for outputting a c. A (nondeterministic) FST for this function is given in Fig. 13; examples of the mapping are given below in (17).

---

[4]We thank Nate Koser for this example.

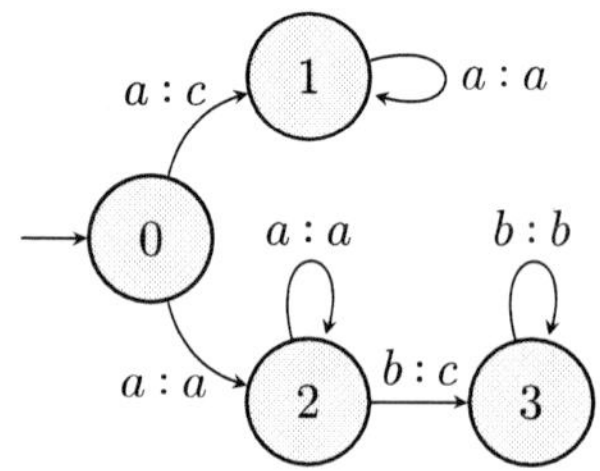

Figure 13: A properly regular function definable using QFLFP with both successor s and predecessor p functions.

$$(17) \quad aaaaaa \mapsto caaaaa$$
$$aababa \mapsto aacaba$$

It can be shown that this function is not subsequential, and thus there is no equivalent SFST for the FST in Fig 13. The reason is made clear by the definition in (16): the transduction looks *ahead* to the right to check if an initial a should be output as a c, but looks *behind* to the left to check if a b is the first b. These cannot both be accomplished by a deterministic FST that reads either right-to-left or left-to-right. Thus, restricting the signature to either p or s is crucial to capturing the behavior of subsequential FSTs.

7 Discussion

We have introduced a class of functions called QFLFP that are defined as graph interpretations using quantifier-free first order logic formulas augmented with a least fixed point operator. In this section we discuss some of the implications of using this class to model phonological processes and highlight a few important areas of future work.

One of the main advantages of such logical characterizations of phonological processes is that they enable a unified approach to both linear and non-linear representations. The graphs used to represent strings in this approach can be extended to represent additional structure used in phonological theory such as autosegmental representations and feature geometry (Goldsmith, 1976; Clements, 1985), syllable constituents (Selkirk, 1984), or metrical structure (Hayes, 1995). This flexibility allows us to directly apply the notion of subsequentiality to other types of representations, which in turn enables more direct comparison among types of phonological phenomena and theories of phonological representation, as we can change the models to accommodate the added structure but maintain the restrictions on the logical formalism (see Chandlee and Jardine, 2019, for an example). This was not as straightforward with finite-state automata.

We have argued that the QFLFP class provides an attractive fit to the observed typology of phonological functions, capturing both local and long-distance phenomena, which the previously defined ISL and OSL classes cannot do without introducing the mechanism of string markup. Compared to the subsequential functions, QFLFP can describe functions dependent on even/odd parity in a local way—such as assigning iterative stress—but not in an unbounded way (e.g., identify every other vowel regardless of how many consonants intervene). The subsequential functions can describe such unbounded even/odd phenomena, which are hypothesized to not exist in phonology. In this way QFLFP appears to be a better fit to the typology compared to the already restrictive subsequential functions.

Future work will aim to identify the limits of QFLFP and the kinds of phonological processes and process interactions that it cannot describe. Such cases will point to the ways the class of functions can be extended and/or modified to provide an even better fit to the range of attested phenomena without including logically possible but unattested patterns. For example, what is the expressivity of QFLFP when two-place predicates (see Koser et al., 2019) or embedded `lfp` operators are allowed?

There are also several theoretical questions left to be addressed. Are QFLFP functions closed under composition? We conjecture that they are not, as per the discussion of the OSL function in Fig. 11 that we conjecture not to be QFLFP. The reason is that functions in which `lfp` operators are embedded in other `lfp` operators appear to be strictly more expressive than those which are not. We also leave an abstract characterization of the QFLFP functions to future work. Such a characterization would lead to a definitive answer to the conjecture that SUBSEQ is a proper superset of QFLFP, as well as whether or not QFLFP is closed under composition. Finally, this paper has considered QFLFP defined over string signatures with only a single ordering function p. The same logic defined over string signatures defined with both p and the successor function s will almost certainly be more expressive, but how much more expressive is an interesting question for future work.

8 Conclusion

QFLFP graph interpretations combine the restrictiveness of quantifier-free first order logic with an operator that can recursively reference the output structure. This allows us to model phenomena beyond the reach of the ISL functions, including iterative spreading processes and long-distance agreement. This class of functions appears to cross-cut several subregular function classes that have previously been applied to the modeling of phonological processes. Because it is still a subset of the subsequential functions, however, it is learnable from positive data. This combination of desirable properties incidates that QFLFP is an important step toward the goal of identifying and understanding the computational nature of phonological processes.

References

Benjamin Ao. 1991. Kikongo nasal harmony and context-sensitive underspecification. *Linguistic Inquiry*, 22(2):193–196.

Jane Chandlee. 2014. *Strictly Local Phonological Processes*. Ph.D. thesis, University of Delaware.

Jane Chandlee, Rémi Eyraud, and Jeffrey Heinz. 2015. Output strictly local functions. In *Proceedings of the 14th Meeting on the Mathematics of Language (MoL 2015)*, pages 112–125, Chicago, USA. Association for Computational Linguistics.

Jane Chandlee and Jeffrey Heinz. 2012. Bounded copying is subsequential: implications for metathesis and reduplication. In *Proceedings of SIGMORPHON 12*.

Jane Chandlee and Jeffrey Heinz. 2018. Strict locality and phonological maps. *LI*, 49:23–60.

Jane Chandlee and Adam Jardine. 2019. Autosegmental strictly local functions. *Transactions of the Association for Computational Linguistics*, 7:157–168.

Jane Chandlee and Steven Lindell. in prep. A logical characterization of strictly local functions. In Jeffrey Heinz, editor, *Doing Computational Phonology*. OUP.

G. N. Clements. 1985. The geometry of phonological features. *Phonology Yearbook*, 2:225–252.

Joost Engelfriet and Hendrik Jan Hoogeboom. 2001. MSO definable string transductions and two-way finite-state transducers. *ACM Transations on Computational Logic*, 2:216–254.

Emmanual Filiot. 2015. Logic-automata connections for transformations. In *Logic and Its Applications (ICLA)*, pages 30–57. Springer.

Emmanuel Filiot and Pierre-Alain Reynier. 2016. Transducers, logic, and algebra for functions of finite words. *ACM SIGLOG News*, 3(3):4–19.

John Goldsmith. 1976. *Autosegmental Phonology*. Ph.D. thesis, Massachussets Institute of Technology.

Bruce Hayes. 1995. *Metrical stress theory*. Chicago: The University of Chicago Press.

Jeffrey Heinz and Regine Lai. 2013. Vowel harmony and subsequentiality. In *Proceedings of the 13th Meeting on the Mathematics of Language (MoL 13)*, pages 52–63.

Mans Hulden. 2009. *Finite-State Machine Construction Methods and Algorithms for Phonology and Morphology*. Ph.D. thesis, University of Arizona.

Adam Jardine. 2016. *Locality and non-linear representations in tonal phonology*. Ph.D. thesis, University of Delaware.

R.M. Kaplan and M. Kay. 1994. Regular models of phonological rule systems. *Computational Linguistics*, (20):371–387.

Nathan Koser, Christopher Oakden, and Adam Jardine. 2019. Tone association and output locality in nonlinear structures. In *Supplemental proceedings of the 2018 Annual Meeting on Phonology*. Linguistics Society of America.

Leonid Libkin. 2004. *Elements of Finite Model Theory*. Berlin: Springer-Verlag.

Huan Luo. 2017. Long-distance consonant agreement and subsequentiality. *Glossa: A Journal of General Linguistics*, 2(1):52.

Mehryar Mohri. 1997. Finite-state transducers in language and speech processing. *Computational Linguistics*, 23(2):269–311.

David Odden. 1994. Adjacency parameters in phonology. *Language*, 70(2):289–330.

J. Oncina, J. García, and E. Vidal. 1993. Learning subsequential transducers for pattern recognition interpretation tasks. *IEEE Transactions on Pattern Analysis and Machine Intelligence*, (15:5):448–457.

Farid M. Onn. 1980. *Aspects of Malay Phonology and Morphology: A Generative Approach*. Kuala Lumpur: Universiti Kebangsaan Malaysia.

Amanda Payne. 2017. All dissimilation is computationally subsequential. *Language: Phonological Analysis*, 93(4):e353–e371.

Marcel Paul Schützenberger. 1977. Sur une variante des fonctions séquentielles. *Theoretical Computer Science*, 4:47–57.

Elisabeth Selkirk. 1984. On the major class features and syllable theory. In Morris Halle, Mark Aronoff, and R.T. Oehrle, editors, *Language sound structure: Studies in phonology*. Cambridge, Mass.: MIT Press.

Some classes of sets of structures definable without quantifiers

James Rogers
Earlham College
jrogers@cs.earlham.edu

Dakotah Lambert
Earlham College
djlambe11@earlham.edu

Abstract

We derive abstract characterizations of the Strictly Piecewise Local (SPL) and Piecewise Locally Testable (PLT) stringsets. These generalize both the Strictly Local/Locally Testable stringsets (SL and LT) and Strictly Piecewise/Piecewise Testable stringsets (SP and PT) in that SPL constraints can be stated in terms of both adjacency and precedence.

We do this in a fully abstract setting which applies to any class of purely relational models that label the points in their domain with some finite labeling alphabet. This includes, for example, labeled trees and graphs. The actual structure of the class of intended models only shows up in interpreting the abstract characterizations of the definable sets in terms of the structure of the models themselves.

1 Introduction

The ultimate goal of this paper is a characterization of the Piecewise Locally Testable and Strictly Piecewise Local classes of sets of strings. But in getting there we employ a very general technique that, with the exception of a single step (the definition of *realizability*) applies to any class of relational structures and yields:

- a quantifier-free logic that is propositional in the sense of "can be interpreted via truth tables as a canonical, but uninteresting, class of models" (but not PC, as in Propositional Calculus),

- an algebraic setting that, modulo the definition of realizability, provides an abstract characterization of the sets of structures definable in that logic, which may or may not be all that useful in itself, but which is strong enough to support more natural characterizations.

The reason that definability with respect to these quantifier-free logics is interesting is that it identifies the sets of structures that are definable purely in terms of the explicit components of the structures themselves, without any auxiliary mechanisms such as distinguishing points in terms independent of their labels (by assigning variables to them, for instance, or associating them with states) or refinements of the label alphabet (by adding features, for instance). This gives a near minimal notion of definability and a class of constraints that can be checked without inferring any information beyond what is explicitly present in the structure itself.

1.1 Overview of the paper

In Section 2 we introduce the Piecewise Local Hierarchy and provide some motivation for exploring its propositional levels. In Section 3 we introduce relational models and their local factors. These are ordinary mathematical models over a purely relational signature which include unary relations that we can interpret as labeling the points in the domain. Beyond that, while the actual structural properties of the models are given by a definition of what counts as an intended model, those properties are inconsequential for nearly all of what follows. As an example we define a class of word models, models for strings that include relations for both successor and precedence.

In Section 4 we introduce a propositional logic based on local factors as atoms and define the class of *Locally Definable* sets of structures as those definable in that logic and the class of *Strictly Locally Definable* sets of structures as those definable by conjunctions of negative literals of that logic. This notion of locality extends that of McNaughton and Papert (1971) to adjacency with respect to any of the non-unary relations of the signature.

In Section 5 we consider sets of factors as models of the logic rather than the relational structures themselves. The function taking a structure to the set of its factors maps models of the first sort (the structures themselves) to models of the second (sets of factors). The advantage of this move is that the space of sets of factors is a finite Boolean Algebra. In this setting it is easy to prove that a set of sets of factors is Strictly Local if and only if (iff) it is a principal ideal in that space. The cost of this move is that not all sets of factors are the image of one of the intended relational structures. Those that are we refer to as being *realizable.*

We then return, in Section 6, to the properties of the Locally and Strictly Locally Definable sets of structures and develop abstract characterizations of these classes. Up until this point, everything we have done applies to any class of relational structures, regardless of the actual structural properties of the intended models (strings, for example, or trees). The characterization of the Local sets is valid for all classes of relational structures, but the last step of the characterization of the Strictly Locally Definable sets depends critically on the notion of realizability. The section ends by completing the characterization for models of strings.

These characterizations can be hard to apply in their fully abstract form. In Section 7 we fix the notion of realizability for models of strings over a signature that includes both successor and precedence, and derive closure properties that are generalizations of the well-known characterizations of the Strictly Local (successor only) and Strictly Piecewise (precedence only) sets of strings.

In Section 8 we consider the learnability of the definable sets of structures. Following that we give both an example of a phonotactic constraint not that is SPL definable and one that separates SPL from both SF and the Tier-based Strictly Local stringset (defined there). We then close with some concluding remarks.

2 The Piecewise-Local Hierarchy

The Piecewise-Local Hierachy (Figure 1) organizes the Local and Piecewise classes of stringsets, introduced in McNaughton and Papert (1971) and extended by Brzozowski and Simon (1973), Simon (1975), Straubing (1985), Thérien and Weiss (1985), Beauquier and Pin (1991) and others, on the basis of model-theoretic definability with respect to word models (see Example 1) along two dimensions: signature (successor alone, less-than alone, or both) and strength of the logical machinery, from the propositional logic discussed below (Section 4) to Monadic Second-Order.

The characterization of Regular stringsets by MSO definability is due to Medvedev (1964), Büchi (1960) and Elgot (1961). This work established the relationship between model-theory of ordered structures and computational structures that spawned the study of Descriptive and Structural Complexity, Finite Model-Theory and other areas of Graph Theory, Abstract Algebra, Theorem Proving and Discrete Math. The characterization of the Star-Free stringsets (SF—definable by regular expressions with complement but not Kleene-closure) by $FO(<)$, First-Order definability with less-than (or both less-than and successor, since successor is FO definable from less-than) is due to McNaughton and Papert (1971), which spawned the work of Brzozowski, Simon, Beauquier and Pin cited above. Thomas (1978) established the characterization of the Locally Threshold Testable (LTT) stringsets by FO definability with successor alone, $FO(+1)$.

Our exploration of the Piecewise-Local hierarchy was motivated by Heinz's exploration of learnability of phonotactic stress patterns (Heinz, 2007). Our research group at Earlham College, over the course of several years, constructed computational tools to classify the patterns in the StressTyp2 (Goedemans et al., 2015) collection of stress patterns that have automata-theoretic semantics, about two-thirds of the 750 lects in the collection, covering a broad range of human languages. These fall into 106 distinct patterns.

Initially, we identified the 82 that are Strictly-Local. In exploring the remainder, we started working with constraints expressed in the propositional logic introduced here in Section 4. Constraints definable using just successor are Locally Testable (LT); Strictly Local (SL) constraints are those that are definable by conjunctions of negative literals. Some constraints, the requirement that every word assigns primary stress to some syllable (obligatoriness) or the requirement that primary stress either falls on a heavy syllable or on the final syllable, while not SL, are clearly the complement of SL constraints (co-SL), disjunctions of positive literals, which share the explicit nature of SL constraints.

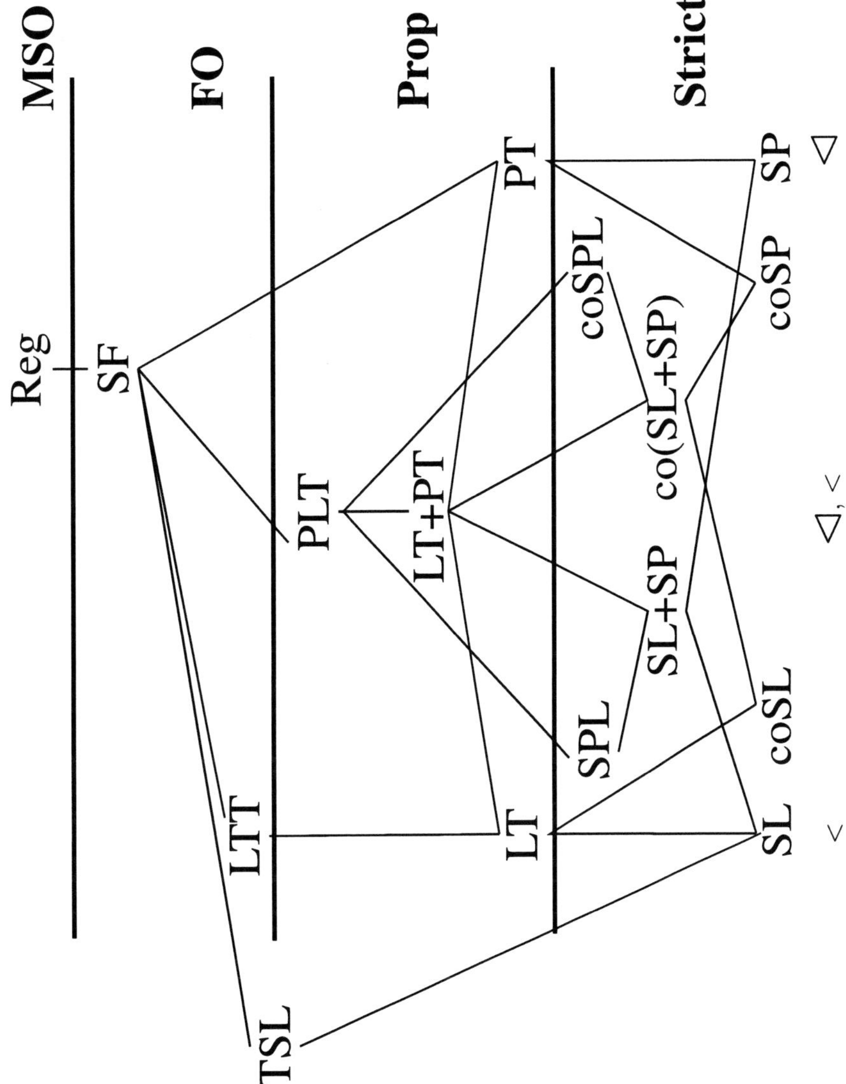

Figure 1: The Piecewise Local Hierarchy

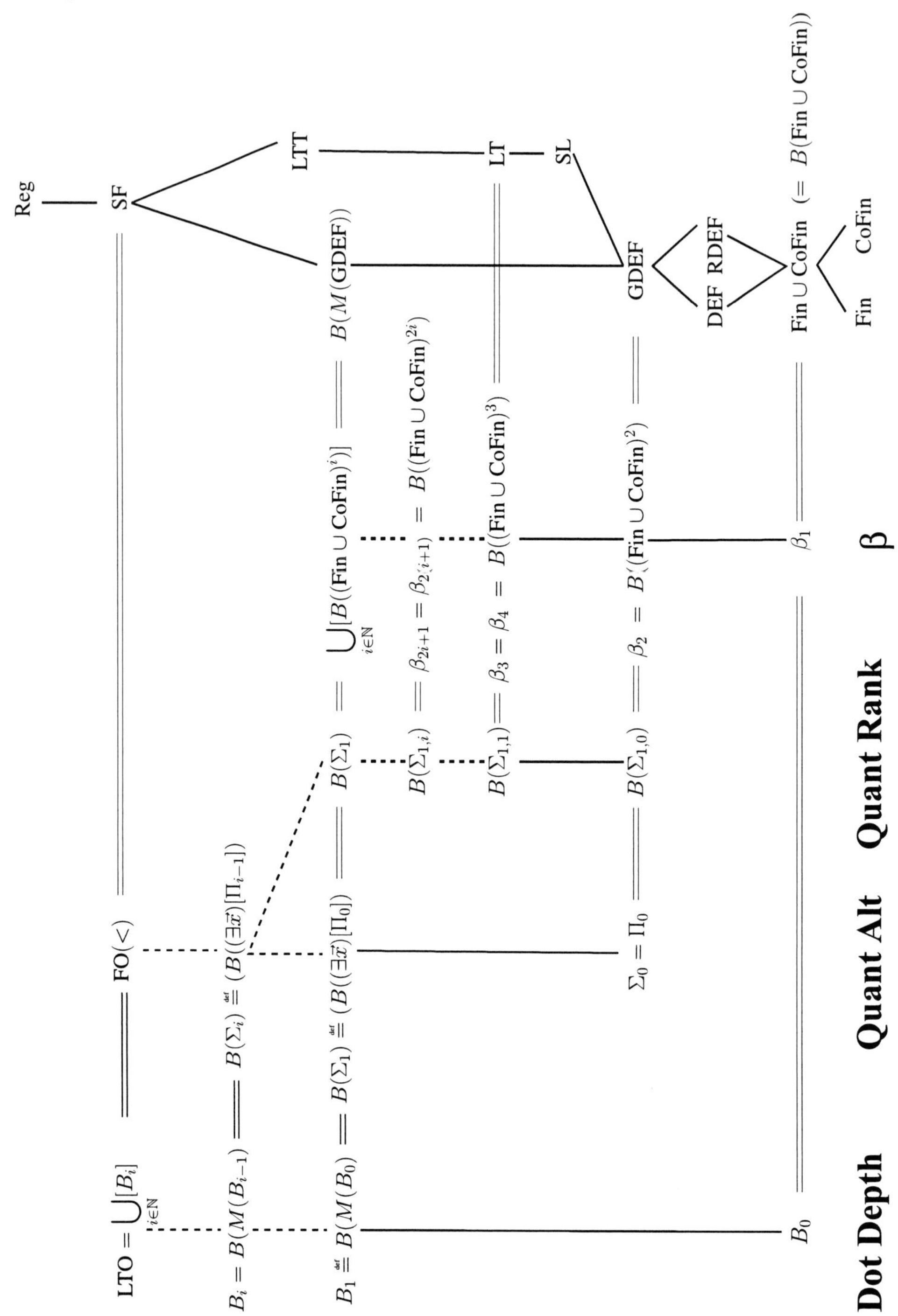

Figure 2: Some Other Local Hierarchies

Some constraints, the requirement that if primary stress falls on a heavy syllable it must be the first heavy syllable, for example, are not even definable in FO($+1$). These are examples of long-distance phonotactics which are amenable to being defined in terms of subsequences (sequences of symbols that occur in order but not necessarily adjacently). These are definable using less-than without the aid of successor; they are Piecewise Testable (PT) constraints. The Piecewise Testable stringsets were introduced by Simon (1975) and are the analog of the LT stringsets based on subsequences rather than substrings. Some, the requirement that primary stress does not fall on more than one syllable (culmanitivity), for example, can be expressed as conjunctions of negative piecewise literals, in SP (Rogers et al., 2010). Obligatoriness, since the only factor involved is a single symbol, can be expressed as the complement of an SP stringset, it is co-SP, as well as co-SL.

Conjunctions of SL, co-SL and SP constraints (SL + co-SL + SP) cover 98 of the 106 patterns in the database. Of the remaining eight, two are properly regular, involving covert alternation. The rest are all of the form: if stress falls on a final syllable that is heavy, then a syllable of some other type (an unstressed heavy, for example) does not occur. While LT they are not expressible as SL or SP constraints or their complement. They are, on the other hand expressible as a negative literal that uses both adjacency (for the identification of the final syllable) and less-than (for the long-distance aspect). This, in addition to the obvious theoretical interest, is what led us to explore the Piecewise Locally (PLT) Testable and Strictly Piecewise Local (SPL) stringsets. Except for those two properly regular patterns, all of the automata-theoretic patterns in StressTyp2 are definable in SPL + co-SL.

2.1 Some Other Sub-Regular Hierarchies

An alternative way of partitioning the Star-Free stringsets is via the dot-depth and β hierarchies (Figure 2). The former is due to Schützenberger (1965) and Brzozowski and Knast (1978). McNaughton and Papert (1971) had already established that the Star-Free stringsets are equal to closure of the Locally Testable stringsets under mixed concatenation and Boolean operations (Locally Testable with Order, LTO). Brzozowski and Knast (1978) establishes an infinite hierarchy, building from the class of finite and co-finite stringsets via concatenation closure (M in the figure) followed by Boolean closure (B), alternately, which partitions LTO. Brzozowski and Simon (1973) refines the dot-depth 1 level into an infinite hierarchy built on concatenation of successively more factors, closed under Boolean operations. The second level (β_2) is equivalent to the class of Generalized Definite stringsets (GDEF, stringsets determined by their initial and final substrings); The third, equivalently fourth, level is equivalent to the LT stringsets.

Thomas (1982), using a somewhat different notion of word model, characterizes these with respect to FO definability. The dot-depth hierarchy (albeit starting at GDEF for Σ_0/Π_0) corresponds to the standard notion of quantifier alternation. The β hierarchy corresponds to Σ_1 stratified by quantifier rank.

The citations given here are a very sparse (and idiosyncratic) sample of the incredibly broad and deep body of work over the last 60 years that has its foundations in those initial results, a testament to the fundamental nature of the results. Perhaps the best route into the theories of word models is the books by McNaughton and Papert and by Straubing (1994).

3 Some definitions

3.1 Relational Models

To be precise about terminology, a relational signature, $\mathbb{R}$, is a ranked alphabet of relation symbols $\{\mathbb{R}^i \mid i \in \mathbb{N}\}$, where the symbols in $\mathbb{R}^i$ represent i-ary relations. Let $\mathbb{R}^*$ be the union of the symbols in $\mathbb{R}^i$ for all i. We assume that $\mathbb{R}$ is finite.

An $\mathbb{R}$-structure is a tuple $\mathcal{A} = \langle A, R_1^{\mathcal{A}}, R_2^{\mathcal{A}}, \ldots \rangle$ where A is the domain and the $R_i^{\mathcal{A}}$ are interpretations of the appropriate arity of symbols chosen from $\mathbb{R}^*$.

Example 1 (Word Models). *Let w be a string over the alphabet Σ. Let $|w|$ be the length of w. A Word Model for w is a structure:*

$$\mathcal{M}_{\Sigma}^{\lhd,<}(w) \stackrel{def}{=} \langle \mathcal{D}^w, \lhd^w, <^w, \rtimes^w, \ltimes^w, P_{\sigma}^w \rangle_{\sigma \in \Sigma}$$

where:

$\mathcal{D}^w$—*is isomorphic to an initial segment $\langle 0, 1, \ldots, |w| + 1 \rangle$ of $\mathbb{N}$ (the Natural numbers).*

$\lhd^w$—*is the successor relation on $\mathcal{D}^w$.*

$<^w$—*is the proper precedence relation on $\mathcal{D}^w$.*

$\rtimes^w$—*is the singleton set containing the minimum position in D^w.*

$\ltimes^w$—*is the singleton containing the maximum position in D^w.*

P_σ^w—*is the set of positions in w at which the symbol σ occurs.*

The sets $\rtimes^w$, $\ltimes^w$ and P_σ^w, for each $\sigma \in \Sigma$ partition D^w (they are pairwise disjoint and their union is D^w).

Let $\mathcal{M}_\Sigma^{\lhd,<}$ denote the class of all word models over Σ.

This definition of a word model differs in certain respects from definitions that may be familiar from prior work. In particular, the endmarkers $\rtimes, \ltimes$ are explicit in the structure and mark points that are adjoined to the ends of the set of positions in the string. Thus, if $w = \langle \sigma_1, \sigma_2, \ldots, \sigma_{|w|} \rangle$, then $\mathbf{card}(D^w) = |w| + 2$ and $i \in P_\sigma^w$ iff $\sigma_i = \sigma$. Also this type of word model includes both the successor and precedence relations. When we look at specific classes in the Piecewise Local sub-regular hierarchy we have, heretofore, employed reducts of this signature including one or the other of the ordering relations, but not both. Here we can obtain the same restrictions by varying parameters restricting their usefulness.

It is important to note that a structure is not necessarily a word model simply because it shares the same signature as these word models. In particular, the interpretations of '$\mathcal{D}^w$', '$\lhd^w$', '$<^w$', '$\rtimes^w$' and '$\ltimes^w$' are not arbitrary, but required to satisfy the axioms of finite discrete linear orders under the usual interpretation of the symbols. We refer to these as the *structural* relations—they form the 'bones' of the intended class of structures—and we refer to those structures that satisfy the axioms as the *intended models*. These notions generalize to other classes of labeled relational structures which exhibit particular structural properties. Word models corresponding to distinct strings differ only in the size of the domain and in the interpretations of the P_σ.[1]

In the core of this paper we temporarily turn to sets of factors (defined in Section 3.3) as mod-

els, which characterize the freely generated structures over the given signature, and encapsulate the theory of the intended structures in the notion of "realizability" (Section 5.3) which picks out the sets of factors that actually correspond to a particular well-formed structure. The core results are valid independent of the definition of realizability, which puts the focus squarely on the defining power of the quantifier-free logic, abstracting away from properties that distinguish a class of intended models from another with the same signature.

Henceforth, when we refer $\mathbb{R}$-structures, we will mean the class of intended structures, however that definition may be restricted. When we discuss the freely generated structures over the signature $\mathbb{R}$, including those that may not be well-formed, we will explicitly say so.

3.2 Homomorphisms and Embeddings

Suppose $\mathcal{A}$ and $\mathcal{B}$ are $\mathbb{R}$-structures. The following definitions are from (Hodges, 1993).

A *homomorphism* from $\mathcal{A}$ to $\mathcal{B}$ is a (total) function $h : A \to B$ such that:

$$R \in \mathbb{R}^i \text{ and } \vec{a} \in R^{\mathcal{A}} \Rightarrow h(\vec{a}) \in R^{\mathcal{B}}.$$

Note that this only requires that the images of the $\vec{a}$ that are in the interpretation of R in $\mathcal{A}$ are included in the interpretation of R in $\mathcal{B}$. It says nothing about other $h(\vec{a})$ that might also be included in $R^{\mathcal{B}}$.

An *embedding* (or strong homomorphism) from $\mathcal{A}$ to $\mathcal{B}$ is a (total) function $h : A \to B$ such that h is a homomorphism that is is strengthened to

$$R \in R^i \text{ and } \vec{a} \in R^{\mathcal{A}} \Leftrightarrow h(\vec{a}) \in R^{\mathcal{B}}$$

We note the difference because "homomorphism" is often taken in the stronger sense, but we necessarily need the weak sense. Otherwise if the image of A in B includes any tuple $h(\vec{a})$ in the domain of the interpretation of a relation R in $\mathcal{B}$, then $\mathcal{A}$ must include the pre-image of $R^{\mathcal{B}}(h(\vec{a}))$ in $R^{\mathcal{A}}$ as well. In this way the interpretation of R in $\mathcal{B}$ would restrict the structure of $\mathcal{A}$.

3.3 Neighborhoods and factors

The next few definitions are based on those in Libkin (2004). The first two are ubiquitous in the Theory of Finite Models.

Definition 1. Let $\mathcal{A}$ be a relational structure as above and $a \in A$. The (domain of the) *r-Ball*

---

[1] Although in this case, the interpretations of the P_σ is not entirely free, either, in that we require those interpretations, along with those of the end markers, to partition the domain. Relaxing that actually leads to a more flexible notion of string-like structures that are useful in many applications.

around a in $\mathcal{A}$ (denoted $B_r^{\mathcal{A}}$) is defined inductively as follows:

$$
\begin{aligned}
B_0^{\mathcal{A}}(a) \;=\; & \{a\} \\
B_{i+1}^{\mathcal{A}}(a) \;=\; & B_i^{\mathcal{A}}(a) \cup \\
& \{a' \mid (\exists R \in \mathbb{R}, \\
& \quad \vec{a} \in R^{\mathcal{A}}, a'' \in B_i^{\mathcal{A}}(a)) \\
& \quad [a', a'' \text{ both occur in } \vec{a}]\}
\end{aligned}
$$

The members of $B_i^{\mathcal{A}}(a)$ are all the members of A that are within distance i of a in the Gaifmann graph of $\mathcal{A}$.

Definition 2 (Connectivity). Let $\mathcal{A}$ be an $\mathbb{R}$ structure.

$\mathcal{A}$ is *k-connected* iff for all $a \in A$, $B_{k-1}^{\mathcal{A}}(a) = A$.

Note that every k-connected domain is also j-connected for all $j > k$.

Definition 3 (k-Factor). Let $\mathcal{A}$ and $\mathcal{B}$ be $\mathbb{R}$ structures.

$\mathcal{A}$ is a *k-factor* of $\mathcal{B}$ iff[2]

$$
\begin{aligned}
& 1: \quad \textbf{card}(A) \leq k \\
& 2: \quad A \text{ is } \textbf{card}(A)\text{-connected} \\
& 3: \quad \exists h : A \rightarrow B, \text{ a homomorphism}
\end{aligned}
$$

N.B. In this definition the set of k-factors of a structure includes all of its j-factors for $j \leq k$.

In general there will be many such h. Each one picks out an occurrence of the factor $\mathcal{A}$ in $\mathcal{B}$.

Let $\mathrm{F}_k(\mathcal{B})$ be the set of isomorphism classes of the set of all k-factors of $\mathcal{B}$. We will ignore the difference between an isomorphism class and a canonical representative of that class, so we can consider this to be a set of structures over an anonymous domain of cardinality k.

Proposition 1. *If $\textbf{card}(B)$ is finite then there are only finitely many distinct (up to isomorphism) k-factors of $\mathcal{B}$.*

Lemma 1. *If $\mathbb{R}$ is relational and finite then the cardinality of*

$$
F_k(\mathbb{R}) \overset{def}{=} \bigcup_{\mathcal{B}} [F_k(\mathcal{B})], \quad \mathcal{B} \text{ an } \mathbb{R}\text{-structure}.
$$

is finite.

---

[2]The term "factor" comes from the well known interpretation of strings in a monoid over concatenation, where the definition is immediate. We adopt this fully general definition in order to extend it to arbitrary relational structures, independent of their specific properties.

Proof Sketch. If $\mathbb{R}$ is relational and finite then there are, up to isomorphism, only finitely many $\mathbb{R}$-structures of cardinality k. This is Exercise 6 (Pg. 10) of Hodges, an easy exercise. ∎

We extend F_k to sets of structures in the standard way, as the union of the sets of k-factors of the structures in the set.

3.4 Supposing $k < n$

Suppose $\mathcal{A}$ is an $\mathbb{R}$-structure, $\textbf{card}(A) = k$ and $R \in \mathbb{R}^n \subseteq \mathbb{R}$, as above, and $k < n$. Then, by the pigeon-hole-principle,

$$
(\forall \vec{a} \in R^{\mathcal{A}})[(\exists a \in A)[\\
a \text{ occurs in at least two places in } \vec{a}]].
$$

Let's say that an n-ary relation is anti-reflexive if no individual occurs more than once in any of its tuples. If all $R \in \mathbb{R}$ are anti-reflexive then, for all $R \in R^{n>k}$, $R^{\mathcal{A}} = \emptyset$.

This is not deep. It just says that in the anti-reflexive case (which will be common) k-factors have nothing to say about relations of arity greater than k.

3.5 Aspects of partial orders

The following is taken, primarily, from MacLane and Birkhoff (1967, 1970). A *partial order* is a set equipped with a partial ordering relation $\sqsubseteq$ that is reflexive, transitive and antisymmetric. If $\sqsubseteq$ is not antisymmetric, then it defines a *quasiorder*.[3]

A *lattice* is a partial order that is closed with respect to two binary operators: a greatest lower bound (meet, $\wedge$) and least upper bound (join, $\vee$). Meets and joins are idempotent, associative and commutative and satisfy the absorption law ($x \wedge (x \vee y) = x = x \vee (x \wedge y)$). If they distribute over each other, then the lattice is distributive.

All finite lattices have a unique minimum element ($\bot$) and a unique maximum element ($\top$). If a lattice has a maximum and minimum element and every element has a complement with respect to these ($x \wedge \overline{x} = \bot$) and ($x \vee \overline{x} = \top$) then it is a complemented lattice. If it is complemented and distributive it is a *Boolean* lattice, equivalently, Boolean algebra. If the lattice is Boolean then every element x has a unique complement $\overline{x}$.

---

[3]In our usage, the relationship between quasiorders and partial orders is analogous to that between preorders and total orders.

An element a of a Boolean algebra is an atom iff $a > \bot$ and there is no b such that $a > b > \bot$.

If a set S is partially ordered by $\sqsubseteq$ and I is a non-empty subset of S that is downward closed ($x \in I$ and $y \sqsubseteq x$ implies $y \in I$) and each pair of elements in I has at least one upper bound in I, then I is an *ideal*. If the ideal includes a unique maximum element a then it is the *principal ideal* generated by a, which we will denote $\mathcal{I}(a)$. S and $\sqsubseteq$ will always be clear from the context.

(Principal) filters are defined dually: upward closed and with lower bounds. We will denote the principal filter generated by a as $\mathcal{F}(a)$.

4 Local and Strictly Local Sets of Structures

Definition 4. Let $\mathbb{R}$ be a relational signature and $\mathcal{G}$ be a subset of $F_k(\mathbb{R})$.

Let $L(\mathcal{G}) \stackrel{\text{def}}{=} \{\mathcal{A} \mid F_k(\mathcal{A}) \subseteq \mathcal{G}\}$.

Then $L(\mathcal{G})$ is a *strictly local set* of $\mathbb{R}$-structures.

A set S of $\mathbb{R}$-structures is a *locally definable set* of $\mathbb{R}$-structures iff it is a Boolean combination of strictly local sets.[4]

4.1 Local Logics

4.1.1 Well-Formed Formulae

Let $\mathbf{wff}_k(\mathbb{R})$ be the set of Boolean formulae in which the atomic formulae are the factors in $F_k(\mathbb{R})$. Usually we can be ambiguous about k, letting it be determined by the formula itself.

4.1.2 Satisfaction with respect to $\mathbb{R}$-structures

Each $\mathbb{R}$-structure provides a valuation of the formulae in $\mathbf{wff}_k(\mathbb{R})$ based on its set of factors: if $f \in F_k(\mathbb{R})$ and $\mathcal{A}$ is an $\mathbb{R}$-structure then

$$\mathcal{A} \models f \stackrel{\text{def}}{\iff} f \in F_k(\mathcal{A}).$$

Let Φ be a set of $\mathbf{wff}(\mathbb{R})$ formulae and $\mathcal{A}$ a $\mathbb{R}$-structure. Then

$$\mathcal{A} \models \Phi \stackrel{\text{def}}{\iff} (\forall \varphi \in \Phi)[\mathcal{A} \models \varphi]$$

and the *models* of Φ is the set

$$\mathbf{Mod}(\Phi) \stackrel{\text{def}}{=} \{\mathcal{A}, \text{ a } \mathbb{R}\text{-structure} \mid \mathcal{A} \models \Phi\}.$$

Φ is *consistent* iff $\mathbf{Mod}(\Phi) \neq \emptyset$.

Let Φ and Ψ be sets of $\mathbf{wff}(\mathbb{R})$ formulae. Φ *entails* (logically implies) Ψ ($\Phi \models \Psi$) iff, by definition, $\mathbf{Mod}(\Phi) \subseteq \mathbf{Mod}(\Psi)$ (i.e., for all $\mathbb{R}$-structures $\mathcal{A}$, $\mathcal{A} \models \Phi \Rightarrow \mathcal{A} \models \Psi$). Φ and Ψ are logically equivalent ($\Phi \equiv \Psi$) iff, by definition, $\Phi \models \Psi$ and $\Psi \models \Phi$.

4.2 Local and Strictly Local Definitions

Let $L = L(\mathcal{G})$ for some $\mathcal{G} \subseteq F_k(\mathbb{R})$ be a k-strictly local set of $\mathbb{R}$-structures. $\mathcal{G}$ is the set of permitted factors; the structures in $L(\mathcal{G})$ may not include any factors but these. Let $\overline{\mathcal{G}} = F_k(\mathbb{R}) - \mathcal{G}$, the set of forbidden factors of L. Since $F_k(\mathbb{R})$ is finite, $\overline{\mathcal{G}}$ is as well. Then L includes all and only those structures that do not include any of the factors in $\overline{\mathcal{G}}$. Formally:

$$L = \mathbf{Mod}\left(\bigwedge_{f \in \overline{\mathcal{G}}} [\neg f]\right).$$

Lemma 2. *A set of $\mathbb{R}$-structures is strictly k-local iff it is the set of models of a conjunction of negative literals of $\mathbf{wff}_k(\mathbb{R})$.*

As usual, we interpret sets of formulae conjunctively, thus a set of $\mathbb{R}$-structures is strictly k-local iff it is $\mathbf{Mod}(\Phi)$ where $\Phi \subseteq \{\neg f \mid f \in F_k(\mathbb{R})\}$.

Lemma 3. *Since a set of structures is local iff it is a Boolean combination of Strictly Locally Definable structures, a set of $\mathbb{R}$-structures is k-Locally Definable if it is $\mathbf{Mod}(\Phi)$ for any $\Phi \subseteq \mathbf{wff}_k(\mathbb{R})$.*

5 Definable sets of subsets of $\mathbf{F}_k(\mathbb{R})$

Consider the space of subsets of $F_k(\mathbb{R})$, partially ordered by subset (this is the *powerset algebra* of $F_k(\mathbb{R})$). It is a Boolean algebra in which $\top$ is $F_k(\mathbb{R})$, $\bot$ is $\emptyset$ and the atoms are the singleton sets of individual factors in $F_k(\mathbb{R})$. We will refer to this space of subsets as $\mathbb{B}_k(\mathbb{R})$. Since $F_k(\mathbb{R})$ is finite, $\mathbb{B}_k(\mathbb{R})$ is as well.

Note that F_k maps $\mathbb{R}$-structures to elements of $\mathbb{B}_k(\mathbb{R})$; it is many-one and generally not onto. While we have restricted our attention to $\mathbb{R}$-structures that are well-formed, those well-formedness properties show up in $\mathbb{B}_k(\mathbb{R})$ only in the structure of the sets of factors. $\mathbb{B}_k(\mathbb{R})$ is the freely generated powerset of the set of k-factors that occur in any well-formed $\mathbb{R}$-structure; those subsets may or may not be in the range of F_k.

---

[4]Note that Locally Definable sets of strings form the classes that are usually referred to as Locally or Piecewise Testable. In McNaughton and Papert (1971), these are specified by sets of permitted initial and final strings of length k, usually $k - 1$ in later work, along with sets of permitted internal strings of length k. In this particular model-theoretic setting the endmarkers obviate the need for three sets of permitted factors, moreover "Testable" is more or less implied and we have, for the most part, replaced it with "Definable". On the other hand, we have not been completely consistent in doing so. This inconsistency should not prove to be overly confusing.

Let $\mathcal{A} \sqsubseteq_k \mathcal{B} \overset{\text{def}}{\iff} \mathrm{F}_k(\mathcal{A}) \subseteq \mathrm{F}_k(\mathcal{B})$. This induces a quasiorder on $\mathbb{R}$-structures, in which two $\mathbb{R}$-structures $\mathcal{A}$ and $\mathcal{B}$ are equivalent with respect to $\sqsubseteq_k$ iff they are logically equivalent with respect to $\mathbf{wff}_k(\mathbb{R})$.

N.B. We denote the order relation of the powerset algebra of $\mathbb{B}_k(\mathbb{R})$ by '$\subseteq$' and from this point on reserve '$\sqsubseteq$' for the quasiorder it induces in the space of $\mathbb{R}$-structures.

We are ultimately interested in the properties of the definable sets in that space of $\mathbb{R}$-structures, but will derive them from the properties of the definable subsets of $\mathbb{B}_k(\mathbb{R})$. One of the advantages of $\mathbb{B}_k(\mathbb{R})$ is that it is finite, while the set of $\mathbb{R}$-structures is infinite. More importantly, it has a simple and regular structure that is independent of the details of the properties of well-formed $\mathbb{R}$-structures.

5.1 Satisfaction with respect to Sets of k-factors

To that end, extend '$\models$' to sets of k-factors in the natural way: $S \subseteq \mathrm{F}_k(\mathbb{R})$ satisfies $f \in \mathrm{F}_k(\mathbb{R})$ iff $f \in S$, with the semantics of the Boolean connectives being defined in the usual way. In order to distinguish definable sets of sets of factors from definable sets of $\mathbb{R}$-structures, we will refer to the sets of sets of k-factors that satisfy a given $\varphi \in \mathbf{wff}_k(\mathbb{R})$ as $\mathbf{Mod}^*(\varphi) = \{S \in \mathbb{B}_k(\mathbb{R}) \mid S \models \varphi\}$.

The semantics of the logical connectives '$\wedge$', '$\vee$' and '$\neg$' correspond directly to the order-theoretic operations '$\wedge$', '$\vee$' and '$\cdot$'. This is, of course, no coincidence.

5.2 Strictly Local Sets of k-factors

Following Lemma 2, a subset of $\mathbb{B}_k(\mathbb{R})$ is strictly local iff it is $\mathbf{Mod}^*(\bigwedge_{f \in \Phi}[\neg f])$, for some $\Phi \subseteq \mathrm{F}_k(\mathbb{R})$.

Note that if $f \in \mathrm{F}_k(\mathbb{R})$ then $\mathbf{Mod}^*(f)$ is the principal filter $\mathcal{F}(f)$ in $\mathbb{B}_k(\mathbb{R})$. Thus:

Lemma 4. *A subset of $\mathbb{B}_k(\mathbb{R})$ is strictly local iff it is the intersection of the complements of a (finite) set of principal filters in $\mathbb{B}_k(\mathbb{R})$.*

Let $\mathbf{S}$ be a strictly local subset of $\mathbb{B}_k(\mathbb{R})$. Since filters are upward-closed, their complements are downward-closed, as is $\mathbf{S}$, the intersection of their complements. The elements of $\mathbf{S}$ are necessarily subsets of $\overline{\Phi}$ (i.e., $\mathrm{F}_k(\mathbb{R}) - \Phi$) and $\overline{\Phi} \in \mathbf{S}$. Thus $\mathbf{S} = \mathcal{I}(\overline{\Phi})$, the principal ideal generated by $\overline{\Phi}$.

Lemma 5. *If $\mathbf{S}$ is a strictly local subset of $\mathbb{B}_k(\mathbb{R})$ then $\mathbf{S}$ is a principal ideal in $\mathbb{B}_k(\mathbb{R})$.*

Let $\mathbf{S}$ be any principal ideal in $\mathbb{B}_k(\mathbb{R})$. Since ideals are downward closed, complements of ideals are necessarily upward closed. Let $\Upsilon(\mathbf{S})$ be the set of minimal elements in $\overline{\mathbf{S}}$. Since $\mathbb{B}_k(\mathbb{R})$ is finite, such minimal elements exist. Since it is a Boolean algebra, each of those elements generates a principal filter in $\mathbb{B}_k(\mathbb{R})$. Then $\overline{\mathbf{S}} = \bigcup_{v \in \Upsilon(S)}[\mathcal{F}(v)]$. Thus, $\mathbf{S} = \bigcap_{v \in \Upsilon(S)}[\overline{\mathcal{F}(v)}]$. Since $\mathbb{B}_k(\mathbb{R})$ is finite, so is Υ, thus $\mathbf{S}$ is strictly local.

Theorem 1. *A subset of $\mathbb{B}_k(\mathbb{R})$ is Strictly Locally Definable iff it is a principal ideal.*

5.3 Realizability

So properties of the strictly local subsets of $\mathbb{B}_k(\mathbb{R})$ are, as promised, extremely simple. What we need now is an abstract characterization of the sets of strictly local sets of $\mathbb{R}$-structures based on these properties.

Some caution is required here, since F_k, as a map between the space of structures and the space of sets of factors, is not onto. The fact that an arbitrary set of factors is a subset of the set of factors of an $\mathbb{R}$-structure $\mathcal{A}$ of the intended type does not necessarily mean that it is the set of factors of a well-formed $\mathbb{R}$-structure—for the word models of Example 1 the factors will need to include both '$\rtimes$' and '$\ltimes$', at least. That type of requirement is not, in general, Strictly-Locally Definable. We have incorporated the properties of the intended models implicitly by considering only well-formed structures in our space of structures. The complexity of defining what it means to be well-formed is a meta-logical issue.

We do know that if L is a k-strictly local set of $\mathbb{R}$-structures then $\mathrm{F}_k(L)$ is a subset of a principal ideal in $\mathbb{B}_k(\mathbb{R})$. Moreover, every $\mathbb{R}$-structure that maps into that ideal is in L. But not every element of that ideal is the image of a well-formed $\mathbb{R}$-structure. Those that are, we refer to as *realizable*.

Definition 5. A subset S of $\mathrm{F}_k(\mathbb{R})$ is *realizable* iff there is some set of well-formed $\mathbb{R}$-structures $\mathbf{A}$ such that $\mathrm{F}_k(\mathbf{A}) = S$.

Every strictly local set of $\mathbb{R}$-structures is the pre-image, under F_k, of the set of realizable elements in a principal ideal in $\mathbb{B}_k(\mathbb{R})$.

6 Structure of the Definable Sets of $\mathbb{R}$-structures

Recall that $\sqsubseteq_k$ is the quasiordering of $\mathbb{R}$-structures that corresponds to $\subseteq$ in $\mathbb{B}_k(\mathbb{R})$.

6.1 A Closure Property of Strictly k-Locally Definable Sets

Since F_k maps every strictly k-local set of $\mathbb{R}$-structures into an downward closed set in $\mathbb{B}_k(\mathbb{R})$ if $\mathcal{A} \in L$, a strictly k-local set of $\mathbb{R}$-structures, and $\mathcal{B} \sqsubseteq_k \mathcal{A}$ then $\mathcal{B} \in L$ as well. So k-strictly local sets are all downward closed under $\sqsubseteq_k$.

But we know much more about $F_k(L)$ than it is downward closed. It is, in fact, a subset of a principal ideal that is generated by some set of k-factors, in particular the $\mathcal{G}$ of Definition 4, and that every realizable subset of $\mathcal{G}$ is the image of some structure in L. So, k-strictly local sets will be closed under any operation that does not increase the set of k-factors of its operands and which preserves realizability.

Lemma 6. *If $\oplus$ is an operation on $\mathbb{R}$-structures such that the set of k-factors of the result is a subset of the union of the sets of k-factors of the operands and which preserves realizability, then every strictly k-local set of $\mathbb{R}$-structures is closed under $\oplus$.*

We will refer to such operations as being *conservative*.

This is a closure condition on strictly k-local sets but not a characterization. The other direction of the characterization depends on the theory of the well-formed $\mathbb{R}$-structures, i.e., on the notion of realizability.

6.2 Characterization of the Local and Strictly Local Sets of Structures

6.2.1 Local Sets

Since $\sqsubseteq_k$ also corresponds to entailment with respect to $\mathbf{wff}_k(\mathbb{R})$, two $\mathbb{R}$-structures are equivalent with respect to $\sqsubseteq_k$ iff they are logically equivalent with respect to $\mathbf{wff}_k(\mathbb{R})$. Thus sets of k-local $\mathbb{R}$-structures cannot break the equivalence classes with respect to $\sqsubseteq_k$.

Even stronger, every such equivalence class is determined by the set of factors of the structures in the class.

Lemma 7. *Let $\equiv_k$ denote equivalence with respect to $\sqsubseteq_k$ and $[\mathcal{A}]_k \overset{def}{=} \{\mathcal{B} \mid \mathcal{A} \equiv_k \mathcal{B}\}$. Then $[\mathcal{A}]_k = \mathbf{Mod}(\bigwedge_{f \in F_k(\mathcal{A})}[f] \wedge \bigwedge_{f \in F_k(\mathbb{R}) - F_k(\mathcal{A})}[\neg f])$*

Theorem 2. *A set of $\mathbb{R}$-structures L is k-local iff whenever $\mathcal{B} \equiv_k \mathcal{A}$ then either both $\mathcal{A}, \mathcal{B} \in L$ or both $\mathcal{A}, \mathcal{B} \notin L$.*

This is a completely general characterization. Every k-local set of $\mathbb{R}$-structures, regardless of the theory of the structures, is the union of a set of equivalence classes with respect to $\equiv_k$.

6.2.2 Strictly Local Sets

Note that, in the space of $\mathbb{R}$-models, the inverse of $\sqsubseteq_k$ is conservative, that is, if L is k-Strictly Piecewise Locally Definable (SPL$_k$), $w \in L$ and $v \sqsubseteq_k w$ then $v \in L$. By definition it does not increase the set of k-factors, and v is trivially realizable. This is very close to a characterization of SPL$_k$, but not quite fully general.

For $f \in F_k(\mathbb{R})$, with mild abuse of notation, let $\mathcal{F}_k^{\sqsubseteq}(f) \overset{def}{=} \{\mathcal{A} \in \mathbb{R} \mid f \in F_k(\mathcal{A})\}$. This is the set of $\mathbb{R}$-models, upper-closed with respect to $\sqsubseteq$, that is generated by f. Similarly, let $\mathcal{F}_k^{\sqsubseteq}(S)$, for $S \subseteq F_k(\mathbb{R})$ be the union of $\mathcal{F}_k^{\sqsubseteq}(f)$ for $f \in S$.

Lemma 8. *Each of the following is a consequence of the preceding statements:*

1. $L \in$ SPL$_k$.

2. $L = \bigcap_{f \in S} \left[\overline{\mathcal{F}_k^{\sqsubseteq}(f)} \right]$, $S \subseteq F_k(\mathbb{R})$, finite.

3. $w \in L$ and $v \sqsubseteq_k w \Rightarrow v \in L$. (L is downward closed with respect to $\sqsubseteq_k$.)

4. $L = \overline{\mathcal{F}_k^{\sqsubseteq}(S)}$, for some $S \subseteq F_k(\mathbb{R})$.

Proof. Each step is nearly immediate. By Lemma 2, $L \in$ SPL$_k \Leftrightarrow L = \mathbf{Mod}(\bigwedge_{f \in \overline{\mathcal{G}}}[\neg f])$, where $\mathcal{G}$ is finite, and each of the $f \in \mathcal{G}$ generates an upper-closed set $\mathcal{F}_k^{\sqsubseteq}(f)$. Since these are upper-closed, their complements are downward closed with respect to $\sqsubseteq_k$, as is their intersection.

To see that 3 implies 4, the complement of L is upper-closed with respect to $\sqsubseteq$. Then S, the set of minimal points in $\overline{L}$ witnesses statement 4. That such a set of minimal points exists is a consequence of the fact that there are no infinite properly descending sequences with respect to $\sqsubseteq$, which itself is a consequence of the finiteness of $\mathbb{B}_k(\mathbb{R})$. $\blacksquare$

The only difference between statements 4 and 2 is the requirement that S be finite. This is where the theory of the well-formed structures comes in. For word models, it is a consequence of Higman's Lemma (Higman, 1952) which says that there are

"

no infinite sequences of strings that are pairwise unrelated by $\sqsubseteq$. For certain classes of tree models, it is a consequence of Kruskal's Tree Theorem (Kruskal, 1960), which is similar.

Theorem 3 (Characterization of Strictly Local Sets of Word Models). *A set of word models is SPL_k iff it is downward closed with respect to $\sqsubseteq_k$.*

7 Strictly Piecewise Local Stringsets

SPL is the class of stringsets corresponding to the class of strictly local word models of Example 1. Since these models are linear, factors can be resolved into blocks of positions connected by '$\lhd$' which are, themselves, connected by '$<$'. In the terminology of the Piecewise Local hierarchy, these are subsequences of substrings. Rather than a single parameter to indicate the size of a factor we use j to denote the maximum number of substrings and k to denote the maximum size of the substrings: $SPL_{j,k}$.

Note that $SPL_{1,k}$ coincides with the well known class of SL_k stringsets, which are all and only those strictly definable in the reduct of our word models that eliminates the precedence relation. And $SPL_{j,1}$ coincides with the SP_j stringsets which are all and only those strictly definable in the reduct of our word models that eliminates successor and the end markers.[5]

In what follows we use $F_{j,k}$ and $\sqsubseteq_{j,k}$ rather than the less precise F_{jk+j-1} and $\sqsubseteq_{jk+j-1}$.

We know, already that $SPL_{j,k}$ sets are closed under $\sqsubseteq_{j,k}$, and that $\mathcal{M}_\Sigma^{\lhd,<}(v) \sqsubseteq_{j,k} \mathcal{M}_\Sigma^{\lhd,<}(w)$ iff $F_{j,k}(\mathcal{M}_\Sigma^{\lhd,<}(v)) \subseteq F_{j,k}(\mathcal{M}_\Sigma^{\lhd,<}(w))$ (modulo realizability), and that, more generally, they are closed under every operation that is conservative in the sense of Lemma 6. What we need is a natural operation on strings that is conservative. That depends on realizability.

7.1 Realizability of sets of $F_{j,k}$ factors

Definition 6 (Minimally Realizable). A set of factors $S \subseteq F_{j,k}(\mathcal{M}_\Sigma^{\lhd,<})$ is *minimally realizable* iff there is a sequence of subsets of S: $q_0 \subsetneq q_1 \subsetneq \cdots \subsetneq q_n \subsetneq q_{n+1}$ such that:[6]

$$q_0 = \{\rtimes\}$$
$$q_{i+1} = F_{j,k}(w_i \cdot \sigma_{i+1}),$$
$$\text{for some } w_i \in \{\rtimes\}\Sigma^*,\ \sigma_{i+1} \in \Sigma$$
$$\text{such that } F_{j,k}(w_i) = q_i$$
$$q_{n+1} = F_{j,k}(w_n \cdot \ltimes),$$
$$\text{for some } w_n \in \{\rtimes\}\Sigma^*$$
$$\text{such that } F_{j,k}(w_n) = q_n$$
$$q_{n+1} = S.$$

In this case the S is the set of $F_{j,k}$-factors of $\mathcal{M}_\Sigma^{\lhd,<}(w)$, where $w = \sigma_1\sigma_2\cdots\sigma_n$, and w is a *minimal witness* that such a well-formed word model exists.

Note that every word model that is equivalent to w with respect to $\sqsubseteq_{j,k}$ and also a witness of the realizability of S but only those that have the same length as w are minimal witnesses.

Every $w \in \Sigma^*$ is a witness of the realizability of $F_{j,k}(\mathcal{M}_\Sigma^{\lhd,<}(w))$. If $|w| \leq |v|$ for every $v \in [w]_{j,k}$ then it is a minimal witness.

Proposition 2. *A subset of $F_{j,k}(\mathcal{M}_\Sigma^{\lhd,<})$ is realizable iff it is the union of a finite set of minimally realizable subsets of $F_{j,k}(\mathcal{M}_\Sigma^{\lhd,<})$.*

7.2 Some closure properties of $SPL_{j,k}$ sets

Using the characterization of Theorem 3 to prove non-definability in $SPL_{j,k}$ can be cumbersome. The following closure conditions, extensions of the characterizations in earlier work on SL_k and SP_j, may be somewhat easier to apply.

Theorem 4 (Generalized Suffix-Substitution Closure). *Suppose L is $SPL_{j,k}$. Then if*

- *$u_1 \cdot x \cdot v_1 \in L$ and $u_2 \cdot x \cdot v_2 \in L$, where $|x| = k - 1$,*

- *and either $F_{j-1,k}(u_1) \subseteq F_{j-1,k}(u_2)$ or $F_{j-1,k}(v_2) \subseteq F_{j-1,k}(v_1)$*

then $u_1 \cdot x \cdot v_2 \in L$.

Proof. From Lemma 6 we know that if substitution of suffixes under these conditions is conservative then $SPL_{j,k}$ is closed with respect to it. To see that it does not increase the set of $F_{j,k}(\mathcal{M}_\Sigma^{\lhd,<})$ note, to begin with, that every $F_{1,k}$ factor in $u_1 \cdot x \cdot v_2$ is also in either $u_1 \cdot x$ or in $x \cdot v_2$, thus in $F_{1,k}(u_1 \cdot x \cdot v_1) \cup F_{1,k}(u_2 \cdot x \cdot v_2)$. Suppose $f_1 \cdot f_2$ is a (j,k)-factor of $u_1 \cdot x \cdot v_2$, and that f_1 is a (i,k)-factor of u_1 and f_2 a $(j-i,k)$-factor of $x \cdot v_2$ for some $i > 1$ (otherwise it is necessarily in either $u_1 \cdot x \cdot v_1$ or $u_2 \cdot x \cdot v_2$). Since, by Definition 3,

---

[5] Since the contiguous blocks of a $(j,1)$-factor are all single symbols the presence of '$\lhd$' is inconsequential for the definable sets. Since $\rtimes < x < \ltimes$ for all positions x in the string, their presence is inconsequential as well.

[6] We denote these subsets as q_i to suggest the connection to a finite state automaton, but we have no need to actually construct such an automaton.

$F_{j-1,k}(w)$ includes $F_{j-i,k}(w)$ for all strings w and $i > 1$, $f_1 \in F_{j-1,k}(u_1)$, $f_2 \in F_{j-1,k}(v_1)$ and $f_1 \cdot f_2 \in F_{j,k}(u_1 \cdot x \cdot v_1)$.

To see that realizability is maintained, note that $u_1 \cdot x \cdot v_2$ is a minimal witness in which the initial segment (up through x) of the sequence of subsets of factors is from the minimal witness for $u_1 x \cdot v_1$ and the final segment (from x on) is from the minimal witness for $u_2 \cdot xv_2$. ∎

Theorem 5 (Generalized Subsequence Closure). *Suppose $w \in L \in SPL_{j,k}$.*
Then if

- *$w = u_1 x_1 v x_2 u_2$, where either $|x_1| = |x_2| = k-1$ or $u_1 = \varepsilon$ and $|x_1| < k-1$ or $u_2 = \varepsilon$ and $|x_2| < k-1$*

- *and $F_{1,k}(x_1 x_2) \subseteq F_{1,k}(x_1 v x_2)$*

then $u_1 x_1 x_2 u_2 \in L$.

Proof. First of all, note that whenever f is in $F_{1,k}(u_1 x_1 x_2 u_2)$ then either $f \in F_{1,k}(u_1 x_1)$ or $f \in F_{1,k}(x_1 x_2)$ or $f \in F_{1,k}(x_2 u_2)$. In each case f is also in $\in F_{1,k}(u_1 x_1 v x_2 u_2)$. Thus the blocks of k consecutive factors in $u_1 x_1 x_2 u_2$ occur in the same order in $u_1 x_1 v x_2 u_2$. Consequently $F_{j,k}(u_1 x_1 x_2 u_2) \subseteq F_{j,k}(u_1 x_1 v x_2 u_2)$.

That this preserves realizability follows from the same reasoning as for Generalized Suffix Substitution Closure. ∎

8 Learnability

Strictly Local, Strictly Piecewise and Strictly Piecewise Local Stringsets were studied in a somewhat different form in Heinz (2007), where they were shown to be learnable in the limit from positive data in the sense of Gold (1967). In Heinz (2010b) he generalizes the learning algorithm to a broad class of stringsets on the based on the notion of string extension.

Let A be a class of objects (factors, for example). A *string extension function* is a total function f, mapping Σ^* to finite subsets of A. Each finite subset of A can be interpreted as a grammar G by letting $L(G) = \{w \in \Sigma^* \mid f(w) \subseteq G\}$. Each string extension function f determines a class of stringsets $\mathcal{L}_f$, the class of all stringsets licensed by subsets of A in the range of f.

Clearly $F_{j,k}$ for word models is a string extension function, with A being the set of all factors of word models, and $\mathcal{L}_{F_{j,k}}$ is the class of Strictly Local, Strictly Piecewise or Strictly Piecewise Local stringsets, depending on j and k. If we take A to be the powerset of the set of all factors of word models then $f(w) \overset{\text{def}}{=} \{F_{j,k}(w)\}$ is one, as well, and $\mathcal{L}_f$ is the class of Locally, Piecewise and Piecewise Locally Definable sets.

A *text* for a stringset L is an enumeration of $L \cup \{\#\}$ in arbitrary order, possibly with repeats. If t is a text, then $t[i]$ denotes the initial segment $t(0) \ldots t(i)$. The learning function ϕ for a string extension function f maps initial segments of a text to finite subsets of A:

$$\phi_f(t[i]) = \begin{cases} \emptyset & \text{if } i = -1 \\ \phi_f(t[i-1]) & \text{if } t(i) = \# \\ \phi_f(t[i-1]) \cup f(t(i)) & \text{otherwise.} \end{cases}$$

For sets of word models, this provides a practical learning algorithm.

More generally, F_k for arbitrary $\mathbb{R}$ structures is an extension function for that class of structures. The issue in those cases is where the the enumeration of members of the set comes from. For non-phonotactic linguistic applications, it essentially requires an annotated sample. If the sample is less than fully characteristic, the learned grammar will undergenerate. On the other hand, in all cases it is useful even if the set is non-PLT. It will learn a set of constraints that define the minimal PLT approximation of that set. For an example of the usefulness of these constraints see Rogers and Lambert (2017, to appear).

9 Some Examples from Phonology

In Section 2 we discussed the automata-theoretic patterns in the StressTyp2 database. The six that are Star-Free and require something more than $SL + co\text{-}SL + SP$ can each be shown to include just one additional LT constraint of the form if stress falls on a final syllable that is heavy, then a syllable of some other type (an unstressed heavy, for example) does not occur. Formally these constraints can be expressed as $\acute{H}\ltimes \rightarrow \neg X$, which is logically equivalent to $\neg(X \wedge \acute{H}\ltimes)$. Since if this fails the X must precede the ultimate $\acute{H}$, we can capture this in $SPL_{2,2}$ with the constraint $\neg(X < \acute{H}\ltimes)$.

Since both SL and SP constraints are expressible as SPL constraints all of these stress patterns, other than the two lects of Arabic, are definable in $SPL + co\text{-}SL$. This is significant from a cognitive perspective because in order to check constraints of these forms a mechanism needs only to attend to factors that actually are present, in isolation, in the input string. (See Rogers et al. (2012) for more on this notion of cognitive complexity.)

9.1 Separating PLT from SF and TSL

In their simplest form (Heinz et al., 2011), Tier-based Strictly Local (TSL) constraints are based on a subset of the input alphabet (the *tier* alphabet) along with strictly local constraints in terms of that alphabet. Operationally, the input string is subjected to an alphabetic homomorphism which erases all symbols except for those in the tier alphabet and the remaining string is checked against the SL constraint. The TSL stringsets are all Star-free, properly include the SL stringsets but are incomparable with the LT, PT and SP stringsets, although the intersection of TSL and SP includes long distance phonotactic patterns derived from asymmetric assimilation processes (Heinz, 2010a).

The canonical separation between TSL and these classes is long distance phonotactic dissimulation patterns. As an example of the application of the closure conditions in Section 7.2, we can show that these patterns are not SPL or even PLT.

Latin liquid dissimulation (LLD): every pair of 'l's is separated by at least one 'r' and every pair of 'r's is separated by at least one 'l':

$$(\forall x, y)[\quad (x < y \wedge l(x) \wedge l(y))$$
$$\rightarrow (\exists z)[x < z \wedge z < y \wedge r(z)]]$$
$$\wedge$$
$$(\forall x, y)[\quad (x < y \wedge r(x) \wedge r(y))$$
$$\rightarrow (\exists z)[x < z \wedge z < y \wedge l(z)]]$$

This definition demonstrates that LLD is SF. It is also TSL based on the tier alphabet $\{l, r\}$ and the constraint $\neg(rr) \wedge \neg(ll)$.

We can demonstrate that it is not $\text{SPL}_{j,k}$ for any j and k using either Generalized Suffix Substitution Closure (GSSC) or Generalized Subsequence Closure (GSSeqC),

9.1.1 Using GSSC

Let
$$w_1 = \rtimes(s^{jk}ls^{jk}r)^{jk} \cdot s^{jk} \cdot ls^{jk}r(s^{jk}ls^{jk}r)\ltimes$$

and
$$w_2 = \rtimes(s^{jk}ls^{jk}r)^{jk}s^{jk}l \cdot s^{jk} \cdot r(s^{jk}ls^{jk}r)\ltimes .$$

Both w_1, $w_2 \in L$, but
$$\rtimes(s^{jk}ls^{jk}r)^{jk} \cdot s^{jk} \cdot r(s^{jk}ls^{jk}r)\ltimes \notin L.$$

Therefore, LLD is not $\text{SPL}_{j,k}$ for any j and k.

9.1.2 Using GSSeqC

Let $w_3 \in L$ be a similar string, divided into $u_1x_1vx_2u_2$ as follows:

$$w_3 = $$
$$\rtimes(s^{jk}ls^{jk}r)^{jk} \cdot s^{k-1} \cdot s^kls^k \cdot s^{k-1} \cdot r(s^{jk}ls^{jk}r)\ltimes$$

Then $|x_1| = |x_2| = k - 1$ and $\text{F}_{1,k}(x_1x_2) \subseteq \text{F}_{1,k}(x_1vx_2)$, but
$$w_4 = u_1x_1x_2u_2$$
$$= \rtimes(s^{jk}ls^{jk}r)^{jk} \cdot s^{k-1} \cdot s^{k-1} \cdot r(s^{jk}ls^{jk}r)\ltimes$$
is not in L.

9.1.3 Using $\equiv_{(j,k)}$.

It is not hard to see that $[\mathcal{W}_3]_{(j,k)} = [\mathcal{W}_4]_{(j,k)}$ (equivalently $\text{F}_{j,k}(\mathcal{W}_3) = \text{F}_{j,k}(\mathcal{W}_4)$), where $\mathcal{W}_3$ and $\mathcal{W}_4$ are word models of w_3 and w_4 equivalently. But w_3 satisfies LLD, while w_4 does not.

10 Conclusion

We have explored the model theory of a type of propositional logic based on factors (connected fragments) of structures defined as labeled purely relational models and given characterizations of the Locally and Strictly Locally Definable sets of these structures. Using those tools, we have derived a characterization of the SPL and PLT definable stringsets, which completes the characterization of the propositional levels of the main sequence of the Piecewise Local hierarchy (See Figure 1).

SPL extends SL and SP by adding, on the one hand, precedence constraints and, on the other, adjacency constraints. The interplay of constraints of these types motivated the original definition of TSL and continues to motivate extensions of the class. But TSL remains incomparable with the sub-Star-Free part of the hierarchy. Ultimately, we hope to find a class of structures that will allow us to incorporate TSL in a natural way.

More importantly, we expect that these model-theoretic tools, when applied to trees and other types of labeled graphs will provide insight into local accounts of autosegmental structures (Jardine, 2017) and other multi-tiered structures as well as model-theoretic accounts of syntactic constraints (e.g. Rogers (1998); Graf (2018)).

Acknowledgments

The authors are indebted to Jeff Heinz, Larry Moss and the anonymous referees for detailed and extremely helpful comments.

References

D. Beauquier and Jean-Eric Pin. 1991. Languages and scanners. *Theoretical Computer Science*, 84:3–21.

J.A. Brzozowski and R. Knast. 1978. The dot-depth hierarchy of star-free languages is infinite. *Journal of Computer and System Sciences*, 16(1):37 – 55.

J.A. Brzozowski and I. Simon. 1973. Characterization of locally testable events. *Discrete Math*, 4:243–271.

J. R. Büchi. 1960. Weak second-order arithmetic and finite automata. *Zeitschrift für mathematische Logik und Grundlagen der Mathematik*, 6:66–92.

Calvin C. Elgot. 1961. Decision problems of finite automata design and related arithmetics. *Transactions of the American Mathematical Society*, 98:21–51.

R. W. Goedemans, Jeffrey Heinz, and Harry van der Hulst. 2015. `http://st2.ullet.net/files/files/st2-v1-archive-0415.tar.gz`. Retrieved 24 Jun 2015.

E.M. Gold. 1967. Language identification in the limit. *Information and Control*, 10:447–474.

Thomas Graf. 2018. Why movement comes for free once you have adjunction. In *Proceedings of CLS 53*, pages 117–137.

Jeffrey Heinz. 2007. *The Inductive Learning of Phonotactic Patterns*. Ph.D. thesis, University of California, Los Angeles.

Jeffrey Heinz. 2010a. Learning long-distance phonotactics. *Linguistic Inquiry*, 41(4):623–661.

Jeffrey Heinz. 2010b. String extension learning. In *Proceedings of the 48th Annual Meeting of the Association for Computational Linguistics*, pages 897–906, Uppsala, Sweden. Association for Computational Linguistics.

Jeffrey Heinz, Chetan Rawal, and Herbert G. Tanner. 2011. Tier-based strictly local constraints for phonology. In *Proceedings of the 49th Annual Meeting of the Association for Computational Linguistics*, pages 58–64, Portland, Oregon, USA. Association for Computational Linguistics.

Graham Higman. 1952. Ordering by divisibility in abstract algebras. *Proceedings of the London Mathematical Society*, s3-2(1):326–336.

Wilfird Hodges. 1993. *Model Theory*. Cambridge University Press, Cambridge, UK.

Adam Jardine. 2017. The local nature of tone-association patterns. *Phonology*, 34:385–405.

J. B. Kruskal. 1960. Well-quasi-ordering, the tree theorem, and Vazsonyi's conjecture. *Transactions of the American Mathematical Society*, 95:210–225.

Leonid Libkin. 2004. *Elements of Finite Model Theory*. Texts in Theoretical Computer Science. Springer, Berlin and New York.

Saunders MacLane and Garrett Birkhoff. 1967, 1970. *Algebra*. Macmillan, New York.

Robert McNaughton and Seymour Papert. 1971. *Counter-Free Automata*. MIT Press.

Yu. T. Medvedev. 1964. On the class of events representable in a finite automaton. In Edward F. Moore, editor, *Sequential Machines; Selected Papers*, pages 215–227. Addison-Wesley. Originally published in Russian in Avtomaty, 1956, 385–401.

James Rogers. 1998. *A Descriptive Approach to Language-Theoretic Complexity*. CSLI Publications, Stanford, CA.

James Rogers, Jeff Heinz, Margaret Fero, Jeremy Hurst, Dakotah Lambert, and Sean Wibel. 2012. Cognitive and sub-regular complexity. In Glyn Morrill and Mark-Jan Nederhof, editors, *Formal Grammar 2012*, volume 8036 of *Lecture Notes in Computer Science*, pages 90–108. Springer.

James Rogers, Jeffrey Heinz, Gil Bailey, Matt Edlefsen, Molly Visscher, David Wellcome, and Sean Wibel. 2010. On languages piecewise testable in the strict sense. In Christian Ebert, Gerhard Jäger, and Jens Michaelis, editors, *The Mathematics of Language: 10th and 11th Biennial Conference, MOL 10, Los Angeles, CA, USA, July 28-30, 2007, and MOL 11, Bielefeld, Germany, August 20-21, 2009, Revised Selected Papers*, pages 255–265. Springer Berlin Heidelberg, Berlin, Heidelberg.

James Rogers and Dakotah Lambert. 2017. Extracting forbidden factors from regular stringsets. In *Proceedings of the 15th Meeting on the Mathematics of Language*, pages 36–46. Association for Computational Linguistics.

James Rogers and Dakotah Lambert. to appear. Extracting subregular constraints from regular stringsets. In press.

M.P. Schützenberger. 1965. On finite monoids having only trivial subgroups. *Information and Control*, 8(2):190 – 194.

Imre Simon. 1975. Piecewise testable events. In *Automata Theory and Formal Languages: 2nd Grammatical Inference conference*, pages 214–222, Berlin. Springer-Verlag.

Howard Straubing. 1985. Finite semigroup varieties of the form $v*d$. *Journal of Pure and Applied Algebra*, 36:53–94.

Howard Straubing. 1994. *Finte Automata, Formal Logic, and Circuit Complexity*. "Birkhäuser".

Wolfgang Thomas. 1978. The theory of successor with and extra predicate. *Mathematische Annalen*, 237:121–232.

Wolfgang Thomas. 1982. Classifying regular events in symbolic logic. *Journal of Computer and Systems Sciences*, 25:360–376.

Denis Thérien and Alex Weiss. 1985. Graph congruences and wreath products. *Journal of Pure and Applied Algebra*, 36:205 – 215.

Efficient Learning of Output Tier-Based Strictly 2-Local Functions

Phillip Burness
University of Ottawa
pburn036@uottawa.ca

Kevin McMullin
University of Ottawa
kevin.mcmullin@uottawa.ca

Abstract

This paper characterizes the Output Tier-based Strictly k-Local ($OTSL_k$) class of string-to-string functions, which are relevant for modeling long-distance phonological processes as input-output maps. After showing that any $OTSL_k$ function can be learned when k and the tier are given, we present a new algorithm that induces the tier itself when $k = 2$ and provably learns any total $OTSL_2$ function in polynomial time and data—the first such learner for any class of tier-based functions.

1 Introduction

In this paper, we investigate the class of Output Tier-based Strictly k-Local ($OTSL_k$) functions. In terms of finite state transducers, $OTSL_k$ functions are those for which the output symbol(s) to be written at each timestep depends on the $k - 1$ most recent symbols on the output tape that belong to the relevant 'tier' (a subset of the output alphabet; Heinz et al., 2011), without regard for any non-tier symbols that might have been written between them or after them. We show that they are learnable when the contents of the tier are provided as input to the learner, and introduce an algorithm that provably and efficiently learns any total OTSL function when $k = 2$.

Recent research investigating the computational properties of phonological patterns observed in natural language has shown that many attested processes can be characterized as Strictly Local (SL) functions (Chandlee, 2014; Chandlee et al., 2014, 2015). That is, the output at any given timestep is dependent on the previous $k - 1$ symbols from either the input string (Input Strictly k-Local; ISL_k) or the output string (Output Strictly k-Local; OSL_k). Multiple characterizations of these classes exist and their properties are well-understood. One important distinction between the two is that non-iterative processes are ISL, whereas processes that apply iteratively to multiple targets are OSL. Moreover, efficient learning algorithms exist for both the ISL and OSL functions. The OSL Function Inference Algorithm (OSLFIA; Chandlee et al., 2015) is of particular importance to this paper, as we will show that many of their theoretical results can be generalized to OTSL functions in a natural way.

Long-distance phonological processes, for which a potentially unbounded number of segments may intervene between the trigger and target without being affected in any way, are neither ISL nor OSL for any value of k. For example, Samala has a long-distance process of sibilant harmony in which an underlying /s/ surfaces as [ʃ] if another [ʃ] appears anywhere later in the word. This is seen when, e.g., the perfective suffix /-waʃ/ is added to a root containing /s/, as in /has-xintila-waʃ/→[haʃxintilawaʃ] 'his former gentile name' (Applegate, 1972). This process can be understood as applying iteratively to multiple targets, as in /s-lu-sisin-waʃ/→[ʃluʃiʃinwaʃ] 'It is all grown awry'. Indeed, it seems that the vast majority of attested long-distance processes are enforced iteratively (Kaplan, 2008; Hansson, 2010). As such, we focus this paper on $OTSL_k$ functions in particular, which generalize the OSL_k class in a way that allows us to model these kinds of long-distance processes. We note that $ITSL_k$ functions can be characterized in a similar way and that the learning strategy outlined below could likely be extended to total $ITSL_2$ functions.

While the notion of a tier has long been incorporated into phonological theory (e.g., Clements, 1980; Goldsmith, 1990; Odden, 1994; Heinz et al., 2011; McMullin, 2016), the range of possible tiers is typically assumed to be available to the learner *a priori*. Each possible tier could, for example, be defined in terms of feature specifications or natural classes of segments (e.g., Hayes and Wilson, 2008). Though algorithms have been developed for inducing a relevant tier from a sample of positive training, their success is limited to phonotactic co-occurrence restrictions. This is true both for constraint-based maximum entropy learners (Gouskova and Gallagher, 2019) as well as for algorithms that learn grammars for Tier-based Strictly Local formal languages (Jardine and Heinz, 2016; Jardine and McMullin, 2017). To our knowledge, the algorithm presented below, which we call the Output Tier-based Strictly 2-Local Function Inference Algorithm (OTSL2FIA), is the first algorithm which learns the relevant tier for transformations of underlying representations (strings of input segments) to surface forms (strings of output segments).

The remainder of this paper is organized as follows. Notation and relevant concepts are presented in Section 2. In Section 3, we define the OTSL functions and characterize them in terms of finite state transducers. In Section 4, we highlight several important properties of $OTSL_2$ functions in particular that can be taken advantage of during learning. All aspects of the learning algorithm, along with the theoretical learning results, are described in Section 5. Section 6 discusses how $OTSL_2$ functions can model various phonological processes and identifies several avenues for future research. Section 7 concludes.

2 Preliminaries

2.1 Strings and sets

Given a set S, we write $\mathtt{card}(S)$ to denote its cardinality. For a string w made of symbols from some alphabet Σ, $|w|$ denotes the length of the string. We write Σ^* to denote all possible strings made from the alphabet Σ, while Σ^n denotes all possible strings made from that alphabet with a length of n, and $\Sigma^{\leq n}$ denotes all such strings with a length up to n. The unique string of length 0 (the empty string) is written as λ. Given two strings u and v, we write $u \cdot v$ to denote their concatenation, but often shorten this to uv when context permits. We write $\mathtt{fac}_k(w)$ to denote all the contiguous substrings of length k (the k-factors) contained in a string w.

We assume a fixed but arbitrary total order $\prec$ over the letters of Σ, an order which we extend to all strings in Σ^* by defining the *length-lexicographical order* (Oncina et al., 1993; Chandlee et al., 2015) as follows. String w_1 occurs length-lexicographically before w_2 (written as $w_1 \triangleleft w_2$) when $|w_1| < |w_2|$ or, if $|w_1| = |w_2|$, when $a_i \prec b_i$ where a_i is the ith letter in w_1, b_i is the ith letter in w_2, and i is the first position on which w_1 and w_2 differ. For example, given $\Sigma = \{a, b\}$ where $a \prec b$, we have $\lambda \triangleleft a \triangleleft b \triangleleft aa \triangleleft ab \triangleleft ba \triangleleft bb \triangleleft aaa$ and so on.

A prefix of some string w is any string u such that $w = ux$ and $x \in \Sigma^*$. Similarly, a suffix of some string w is any string u such that $w = xu$ and $x \in \Sigma^*$. Note that any string is a prefix and suffix of itself, and that λ is a prefix and suffix of every string. When $|w| \geq n$, $\mathtt{Pref}^n(w)$ and $\mathtt{Suff}^n(w)$ denote the unique prefix and suffix of w with a length of n; when $|w| < n$, they simply denote w itself. We write $\mathtt{Pref}^*(w)$ to denote the set of all prefixes of w. Also, $\mathtt{Suff}^n(\mathtt{Suff}^n(w_1)w_2)) = \mathtt{Suff}^n(w_1 w_2)$. Given a string w, one of its prefixes p, and one of its prefixes s, we write $p^{-1} \cdot w$ to represent the string w without that prefix p and write $w \cdot s^{-1}$ to represent the string w without that suffix s. For example, $a^{-1} \cdot aba = ba$ and $aba \cdot a^{-1} = ab$. Finally, given a set of strings S, we write $\mathtt{lcp}(S)$ to denote the *longest common prefix* of S, which is the string u such that u is a prefix of every $w \in S$, and there exists no other string v such that $|v| > |u|$ and v is also a prefix of every $w \in S$.

2.2 Functions and transducers

This paper deals exclusively with string-to-string functions, relations that pair every $w \in \Sigma^*$ with at most one $y \in \Delta^*$, where Σ and Δ are the input alphabet and output alphabet respectively. The input language and output language of such a function are $\mathtt{pre\_image}(f) = \{x \mid (\exists y)[x \mapsto_f y]\}$ and $\mathtt{image}(f) = \{y \mid (\exists x)[x \mapsto_f y]\}$, respectively. An important concept is that of the *tails* of an input string w with respect to a function f.

Definition 1. (Tails) *Given a function f and an input $w \in \Sigma^*$, $\mathtt{tails}_f(w) = \{(y, v) \mid f(wy) = uv \wedge u = \mathtt{lcp}(f(w\Sigma^*))\}$.*

In words, $\mathtt{tails}_f(w)$ pairs every possible string $y \in \Sigma^*$ with the portion of $f(wy)$ that is directly attributable to y. That is, it describes the effect that w has on the output of any subsequent string of input symbols. When $\mathtt{tails}_f(w_1) = \mathtt{tails}_f(w_2)$ we say that w_1 and w_2 are *tail-equivalent* with respect to f.

A related concept to tails and tail-equivalency

is the *contribution* of a symbol $a \in \Sigma$ relative to a string $w \in \Sigma^*$ with respect to a function f.

Definition 2. (Contribution) *Given a function f, some $a \in \Sigma$, and some $w \in \Sigma^*$,* $\mathrm{cont}_f(a, w) = \mathrm{lcp}(f(w\Sigma^*))^{-1} \cdot \mathrm{lcp}(f(wa\Sigma^*))$.

In words, for an input string x that has the prefix wa, the contribution of the a in wa is the portion of $f(x)$ that is uniquely and directly attributable to that instance of a.

The Output Tier-based Strictly Local functions that will be introduced below are a proper subclass of the subsequential functions. Oncina and García (1991) show that when a function is subsequential, tail-equivalency will partition Σ^* into finitely many blocks, allowing us to construct a finite-state transducer that computes f. In this paper we use delimited subsequential finite state transducers (DSFSTs; see Jardine et al., 2014), to characterize the class of Output Strictly Local (OSL) functions. The following definition is drawn directly from Chandlee et al. (2015).

Definition 3. *A delimited subsequential finite state trasnducer (DSFST) is a 6-tuple $\langle Q, q_0, q_f, \Sigma, \Delta, \delta \rangle$ where Q is a finite set of states, $q_0 \in Q$ is the unique initial state, $q_f \in Q$ is the unique final state, Σ is the finite input alphabet, Δ is the finite output alphabet, and $\delta \subseteq Q \times (\Sigma \cup \{\rtimes, \ltimes\}) \times \Delta^* \times Q$ is the transition function (where $\rtimes \notin \Sigma$ indicates the start of the input and $\ltimes \notin \Sigma$ indicates the end of the input), and the following hold:*

1. *if $(q, a, u, q') \in \delta$ then $q \neq q_f$ and $q' \neq q_0$*

2. *if $(q, a, u, q_f) \in \delta$ then $a = \ltimes$ and $q \neq q_0$*

3. *if $(q_0, a, u, q') \in \delta$ then $a = \rtimes$ and if $(q, \rtimes, u, q') \in \delta$ then $q = q_0$*

4. *if $(q, a, u, q'), (q, a, u', q'') \in \delta$ then $q' = q''$ and $u = u'$*

Each transition $(q, a, u, q') \in \delta$ can be seen as an instruction to append u to the end of the output tape and to move to state q' upon reading a while in state q. This transition function may be partial, and its recursive extension δ^* is the smallest set containing δ closed under the following conditions: $(q, \lambda, \lambda, q) \in \delta^*$, and $(q, w, u, q'), (q', a, v, q'') \in \delta^* \Rightarrow (q, wa, uv, q'') \in \delta^*$. The initial state of a DSFST has no incoming transitions and has exactly one outgoing transition, which will be for the input $\rtimes$ and does not land in the final state. Furthermore, the final state of a DSFST has no outgoing transitions, and every transition into the final state is for the input $\ltimes$. DSFSTs are also deterministic on the input, such that each state has at most one outgoing transition per input symbol.

The size of a DSFST $\mathcal{T} = \langle Q, q_0, q_f, \Sigma, \Delta, \delta \rangle$ is $|\mathcal{T}| = \mathrm{card}(Q) + \mathrm{card}(\delta) + \sum_{(q,a,u,q') \in \delta} |u|$, and the relation defined by a DSFST is $\mathcal{R}(\mathcal{T}) = \{(x, y) \in \Sigma^* \times \Delta^* \mid (q_0, \rtimes x \ltimes, y, q_f) \in \delta^*\}$.

The DSFSTs we will use below have a special property known as *onwardness*, which informally means that the writing of the output is never delayed. The following formal definition of onwardness and a related lemma are borrowed from Chandlee et al. (2015).

Definition 4. (Onwardness) *A DSFST is* onward *if for every $w \in \Sigma^*$ and $u \in \Delta^*$, $(q_0, \rtimes w, u, q) \in \delta^* \Longleftrightarrow u = \mathrm{lcp}(f(w\Sigma^*))$.*

Lemma 1. *Let the outputs of the edges out of state q be $\mathrm{Outputs}(q) = \{u \mid (\exists a \in \Sigma \cup \{\rtimes, \ltimes\}) \, (\exists q' \in Q) \, [(q, a, u, q') \in \delta]\}$. If $\mathcal{T}$ is an onward DSFST and recognizes f, then $\forall q \neq q_0$, $\mathrm{lcp}(\mathrm{Outputs}(q)) = \lambda$ and $\mathrm{lcp}(\mathrm{Outputs}(q_0)) = \mathrm{lcp}(f(\Sigma^*))$.*

Below we will frequently make reference to the length-lexicographically earliest input string that can lead to a state q in a given transducer $\mathcal{T}$, which we will denote as w_q. A formal definition is provided here for reference.

Definition 5. (Earliest string) *Given a transducer $\mathcal{T} = \langle Q, q_0, q_f, \Sigma, \Delta, \delta \rangle$, the earliest string that leads to $q \in Q$ is $w_q = \min_{\triangleleft} \{w \in \Sigma^* \mid \exists u, (q_0, \rtimes w, u, q) \in \delta^*\}$.*

A distinction that will be important throughout the rest of this paper is that between the writing that occurs in a DSFST as it is reading letters from Σ, and the writing that occurs at the very end (when the DSFST reads $\ltimes$). To make this distinction, Chandlee et al. (2015) defined the prefix function f^p associated with a subsequential function f as follows.

Definition 6. (Prefix function) *Given a subsequential function f, its associated prefix function f^p is such that $f^p(w) = \mathrm{lcp}(f(w\Sigma^*))$.*

Remark 1. *Given a subsequential function f, some $a \in \Sigma$, and some input string $w \in \Sigma^*$, $\mathrm{cont}_f(a, w) = f^p(w)^{-1} \cdot f^p(wa)$ and $\mathrm{cont}_f(\ltimes, w) = f^p(w)^{-1} \cdot f(w)$.*

2.3 Strict locality and tiers

Chandlee (2014) and Chandlee et al. (2014) originally introduced the Input Strictly Local (ISL) and Output Strictly Local (OSL) functions, both of which generalize Strictly Local (SL) stringsets to functions based on one of the defining properties of SL languages, the Suffix Substitution Closure (Rogers and Pullum, 2011). The definitions of the ISL and OSL functions exploit a corollary of this defining property, which Chandlee et al. (2015) call Suffix-defined Residuals. For reasons of space, we only discuss the OSL functions below.

Theorem 1. (Suffix Substitution Closure) *A language L is SL if for all strings u_1, v_1, u_2, v_2 there exists a natural number k such that for any string x of length $k - 1$, if $u_1 x v_1, u_2 x v_2 \in L$, then $u_1 x v_2 \in L$.*

Corollary 1. (Suffix-defined Residuals) *A language L is SL if for all $w_1, w_2 \in \Sigma^*$, there exists a natural number k such that if $Suff^{k-1}(w_1) = Suff^{k-1}(w_2)$ then $\{v \mid w_1 v \in L\} = \{v \mid w_2 v \in L\}$, that is w_1 and w_2 have the same residuals (tails) with respect to L.*

Definition 7. (Output Strictly Local functions) *A function f is OSL_k if for all $w_1, w_2 \in \Sigma^*$, $\mathtt{Suff}^{k-1}(f^p(w_1)) = \mathtt{Suff}^{k-1}(f^p(w_2)) \Rightarrow \mathtt{tails}_f(w_1) = \mathtt{tails}_f(w_2)$.*

Chandlee (2014) and Chandlee et al. (2014, 2015) show that most iterative phonological processes can be modelled with an OSL function, with an important exception being long-distance iterative processes like consonant harmony. This is parallel to the fact that long-distance phonotactics cannot be represented with an SL stringset, which motivated Heinz et al. (2011) to define the Tier-based Strictly Local (TSL) languages—stringsets that are SL after an erasure function has applied, masking all symbols that are irrelevant to the restrictions that the language places on its strings.

Definition 8. (Erasure function) *Given an alphabet Σ, a tier $\Theta \subseteq \Sigma$, and a string $w = a_1...a_n$, $\mathtt{Erase}_\Theta(w) = b_1...b_n$ where for all $i \leq n$, $b_i = a_i$ if $a_i \in \Theta$, else $b_i = \lambda$.*

Informally, $\mathtt{Erase}_\Theta(w)$ returns the string w with all non-tier elements removed. For convenience, we will write $\mathtt{Suff}^n_\Theta(w)$ to mean $\mathtt{Suff}^n(\mathtt{Erase}_\Theta(w))$ in what follows.

Definition 9. (Tier-based Strictly Local languages) *A language L is Tier-based Strictly k-Local (TSL_k) if there is a tier $\Theta \subseteq \Sigma$ and a subset $S \subseteq \mathtt{fac}^k(\rtimes \Theta^* \ltimes)$ such that:*
$$L = \{w \in \Sigma^* \mid \mathtt{fac}^k(\rtimes Erase_\Theta(w) \ltimes) \subseteq S\}$$

3 Output Tier-based Strictly Local functions and transducers

In this section, we define the OTSL functions, which generalize the TSL stringsets to functions in the same way that the OSL functions generalize SL stringsets to functions (see Chandlee, 2014; Chandlee et al., 2015).

Definition 10. (Output Tier-based Strictly Local functions) *A function f is $OTSL_k$ if there is a tier $\Theta \subseteq \Delta$ such that for all $w_1, w_2 \in \Sigma^*$, $\mathtt{Suff}^{k-1}_\Theta(f^p(w_1)) = \mathtt{Suff}^{k-1}_\Theta(f^p(w_2)) \Rightarrow \mathtt{tails}_f(w_1) = \mathtt{tails}_f(w_2)$.*

The OTSL class properly contains the OSL functions, since every OSL_k function can be described as an $OTSL_k$ function whose tier is equal to the entire output alphabet. Note that it is possible for a single OTSL function to be described with more than one tier. For example, the identity function (where $\Sigma = \Delta$ and $f(w) = w$) can be described with any subset of Δ as its tier. We use the term k-tier to describe a tier Θ for which f is $OTSL_k$.

Like the OSL_k functions, the $OTSL_k$ functions can be characterized in automata-theoretic terms. First, we define $OTSL_k$ finite state transducers as follows.

Definition 11. *An onward DSFST $\mathcal{T} = \langle Q, q_0, q_f, \Sigma, \Delta, \delta \rangle$ is $OTSL_k$ for the tier $\Theta \subseteq \Delta$ if:*

1. *$Q = S \cup \{q_0, q_f\}$ with $S \subseteq \Theta^{\leq k-1}$*

2. *$(\forall u \in \Delta^*)$*
   $[(q_0, \rtimes, u, q') \in \delta \Rightarrow q' = \mathtt{Suff}^{k-1}_\Theta(u)]$

3. *$(\forall q \in Q - \{q_0\}, \forall a \in \Sigma, \forall u \in \Delta^*)$*
   $[(q, a, u, q') \in \delta \Rightarrow q' = \mathtt{Suff}^{k-1}_\Theta(qu)]$.

Lemmas 2 and 3, together with Theorem 2, show that the $OTSL_k$ functions and the functions represented by $OTSL_k$ transducers exactly correspond.

Lemma 2. *Let $\mathcal{T} = \langle Q, q_0, q_f, \Sigma, \Delta, \delta \rangle$ be an $OTSL_k$ transducer for the tier Θ. The following holds: $(q_0, \rtimes w, u, q) \in \delta^* \Rightarrow q = \mathtt{Suff}^{k-1}_\Theta(u)$.*

Lemma 3. *Any OTSL$_k$ transducer corresponds to an OTSL$_k$ function.*

Theorem 2. *Given an OTSL$_k$ function f and one of its k-tiers Θ, the DSFST $\mathcal{T} = \langle Q, q_0, q_f, \Sigma, \Delta, \delta \rangle$ defined as follows computes f:*

1. $Q = S \cup \{q_0, q_f\}$ with $S \subseteq \Theta^{\leq k-1}$

2. $(q_0, \rtimes, u, \mathtt{Suff}_\Theta^{k-1}(u)) \in \delta \Longleftrightarrow u = f^p(\lambda)$

*3. $a \in \Sigma, (q, a, u, \mathtt{Suff}_\Theta^1(qu)) \in \delta \Longleftrightarrow$
   $(\exists w)\, [f^p(w) = vr \wedge \mathtt{Suff}_\Theta^{k-1}(vr) = q \wedge$
   $f^p(wa) = vru]$
   where $r = t_1 x_1 t_2 x_2 ... t_{k-1} x_{k-1}$,
   $t_i \in \Theta, x_i \in (\Delta - \Theta)^*$ and $v = f^p(w) \cdot r^{-1}$*

4. $(q, \ltimes, u, q_f) \in \delta \Longleftrightarrow u = f^p(w_q)^{-1} \cdot f(w_q)$

We note that these are trivial extensions of Lemmas 3, 4, and Theorem 2 in Chandlee et al. (2015). Indeed, only two minor changes are necessary for this generalization to OTSL$_k$ functions. First, each instance of $\mathtt{Suff}^{k-1}$ must be replaced with $\mathtt{Suff}_\Theta^{k-1}$. Second, in order to account for the fact that non-tier elements may come between relevant tier elements, certain references to a string $q = t_1 t_2 ... t_{k-1}$ must be rewritten as $r = t_1 x_1 t_2 x_2 ... t_{k-1} x_{k-1}$, where $t_i \in \Theta$ and $x_i \in (\Delta - \Theta)^*$. As the proofs are otherwise identical in structure to those found in Chandlee et al. (2015), we do not provide them here.

It is therefore the case that any OTSL$_k$ function can be represented by an OTSL$_k$ transducer. Informally, this will be an onward DSFST in which the non-initial and non-final states represent the most recent $k-1$ tier symbols written thus far, meaning that this is the only information that will dictate what the DSFST writes upon reading the next input symbol.

As an example, Figure 1 presents an OTSL$_2$ transducer that models the unbounded sibilant harmony in Samala from Section 1. Note that in order to achieve the regressive directionality of the process, we assume that this transducer reads input strings from right-to-left (following, e.g., Heinz and Lai, 2013; Chandlee et al., 2015). Directionality will be further discussed in Section 6.

4 Useful properties of OTSL$_2$ functions

The main goal of this paper is to demonstrate how OTSL functions can be learned from positive data, even without prior knowledge of the tier itself. We note that the tier-induction strategy adopted below

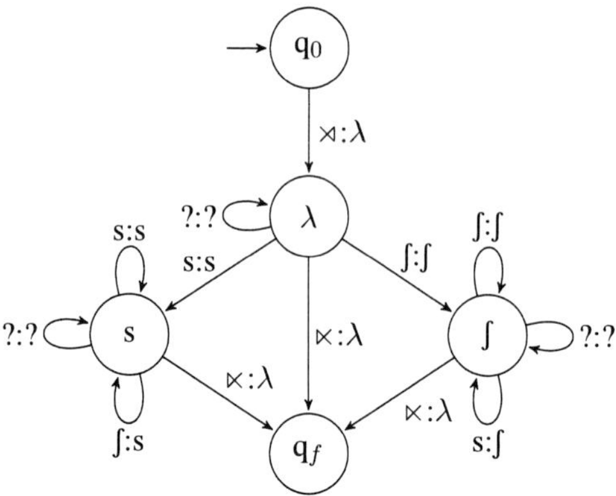

Figure 1: An OTSL$_2$ transducer that models unbounded sibilant harmony, where ? represents any symbol that is not s or ʃ

relies on certain properties that hold when $k = 2$, but not necessarily for greater values of k. These are outlined below.

First, when an OTSL$_2$ function f can be described with more than one 2-tier, the union of any two or more such 2-tiers is also a 2-tier for f.

Lemma 4. *Given an OTSL$_2$ function f, if $\Theta_1 \subseteq \Delta$ and $\Theta_2 \subseteq \Delta$ are both 2-tiers for f, then $\Omega = \Theta_1 \cup \Theta_2$ is also a 2-tier for f.*

Proof. If $\mathtt{Suff}_\Omega^1(f^p(w_1)) = \mathtt{Suff}_\Omega^1(f^p(w_2)) = t$, then $t \in \Theta_1$ or $t \in \Theta_2$. If $t \in \Theta_1$, then $\mathtt{Suff}_{\Theta_1}^1(f^p(w_1)) = \mathtt{Suff}_{\Theta_1}^1(f^p(w_2)) = t \Rightarrow \mathtt{tails}_f(w_1) = \mathtt{tails}_f(w_2)$. If $t \in \Theta_2$, then $\mathtt{Suff}_{\Theta_2}^1(f^p(w_1)) = \mathtt{Suff}_{\Theta_2}^1(f^p(w_2)) = t \Rightarrow \mathtt{tails}_f(w_1) = \mathtt{tails}_f(w_2)$. Therefore, $\mathtt{Suff}_\Omega^1(f^p(w_1)) = \mathtt{Suff}_\Omega^1(f^p(w_2)) = t \Rightarrow \mathtt{tails}_f(w_1) = \mathtt{tails}_f(w_2)$ $\square$

It is this property that allows us to identify a unique *target* tier for an OTSL$_2$ function, which the algorithm can find by flagging and removing elements of Δ from its hypothesis when evidence is found that they cannot be on a relevant tier. We define this canonical 2-tier as follows.

Definition 12. (Canonical 2-tier) *Given an OTSL$_2$ function f, Θ is the canonical 2-tier for f iff there is no other 2-tier $\Omega \subseteq \Delta$ for f such that $\mathtt{card}(\Omega) \geq \mathtt{card}(\Theta)$.*

Remark 2. *Given an OTSL$_2$ function f, its canonical 2-tier Θ is a superset of any 2-tier for f. (This follows immediately from Lemma 4.)*

There is therefore a unique canonical 2-tier (i.e., the largest one) for each OTSL$_2$ function. Interestingly, this can be exploited during the learning process, since it leads to the following useful property of $OTSL_2$ functions.

Lemma 5. *Let $f : \Sigma^* \to \Delta^*$ be an $OTSL_2$ function, let Θ be its canonical 2-tier, and let Ω be such that $\Theta \subset \Omega \subseteq \Delta$. We have $\exists a \in (\Omega - \Theta)$, $\exists w_1, w_2 \in \Sigma^*$, and $\exists x \in \Sigma \cup \{\ltimes\}$ such that $\mathrm{Suff}_\Omega^1(f^p(w_1)) = \mathrm{Suff}_\Omega^1(f^p(w_2)) = a$ and $\mathrm{cont}_f(x, w_1) \neq \mathrm{cont}_f(x, w_2)$.*

Proof. By contradiction. Suppose that the lemma is false. This means that $\forall a \in (\Omega - \Theta)$, $\forall w_1, w_2 \in \Sigma^*$, and $\forall x \in \Sigma \cup \{\ltimes\}$, we have $\mathrm{Suff}_\Omega^1(f^p(w_1)) = \mathrm{Suff}_\Omega^1(f^p(w_2)) = a \Rightarrow \mathrm{cont}_f(x, w_1) = \mathrm{cont}_f(x, w_2)$. Now, since Θ is a 2-tier for f, it is also the case that $\forall b \in \Theta$, $\forall w_1, w_2 \in \Sigma^*$, and $\forall x \in \Sigma \cup \{\ltimes\}$, we have $\mathrm{Suff}_\Omega^1(f^p(w_1)) = \mathrm{Suff}_\Omega^1(f^p(w_2)) = b \Rightarrow \mathrm{cont}_f(x, w_1) = \mathrm{cont}_f(x, w_2)$. Together these imply that $\forall c \in \Omega$, $\forall w_1, w_2 \in \Sigma^*$, and $\forall x \in \Sigma \cup \{\ltimes\}$, we have $\mathrm{Suff}_\Omega^1(f^p(w_1)) = \mathrm{Suff}_\Omega^1(f^p(w_2)) = c \Rightarrow \mathrm{cont}_f(x, w_1) = \mathrm{cont}_f(x, w_2)$.

Since $[\mathrm{Suff}_\Omega^1(f^p(w_1)) = \mathrm{Suff}_\Omega^1(f^p(w_2))$ and $\mathrm{cont}_f(x, w_1) = \mathrm{cont}_f(x, w_2)] \Rightarrow \mathrm{Suff}_\Omega^1(f^p(w_1 x)) = \mathrm{Suff}_\Omega^1(f^p(w_2 x))$, we also have $\mathrm{cont}_f(y, w_1 x) = \mathrm{cont}_f(y, w_2 x)$ for all $y \in \Sigma \cup \{\ltimes\}$. This applies recursively, giving us $\mathrm{Suff}_\Omega^1(f^p(w_1)) = \mathrm{Suff}_\Omega^1(f^p(w_2)) \Rightarrow \mathrm{tails}_f(w_1) = \mathrm{tails}_f(w_2)$, which means that Ω is a 2-tier for f. However, $\mathrm{card}(\Omega) > \mathrm{card}(\Theta)$, contradicting the fact that Θ is the canonical 2-tier for f. $\qquad\square$

Importantly, it follows from Lemma 5 that for any set Ω which is a strict superset of Θ (the canonical 2-tier), we will always be able to find evidence that some member of Ω could not be a member of any 2-tier for f. It is this property of $OTSL_2$ functions that our algorithm makes use of to determine which output symbols are in Θ. Once again, when $k > 2$, this property does not necessarily hold.[1] Accordingly, we restrict ourselves to $k = 2$ when discussing the learning of OTSL functions without prior knowledge of the tier. While $OTSL_2$ functions seem sufficient for modelling a wide range of long-distance phonological processes, we discuss certain exceptions in Section 6.

---

[1] For example, if $\Delta = \{a, b, c\}$, there could be an $OTSL_3$ function for which $\Theta_1 = \{a, b\}$ and $\Theta_2 = \{a, c\}$ are both 3-tiers, but $\Omega = \{a, b, c\}$ is not.

5 Learning OTSL functions

5.1 Learning paradigm

We adopt the criterion for successful learning that requires exact identification in the limit from positive data (Gold, 1967), with polynomial bounds on time and data (de la Higuera, 1997). We first define what it means for a class of functions to be represented by a class of representations.

Definition 13. *A class $\mathbb{T}$ of functions is represented by a class $\mathbb{R}$ of representations if every $r \in \mathbb{R}$ is of finite size and there is a total and surjective naming function $\mathcal{L} : \mathbb{R} \to \mathbb{T}$ such that $\mathcal{L}(r) = t$ if and only if for all $w \in \mathtt{pre\_image}(t), r(w) = t(w)$, where $r(w)$ is the output produced by r given the input w.*

The notions of a sample and a learning algorithm are defined as follows.

Definition 14. (Sample) *A sample S for a function $t \in \mathbb{T}$ is a finite set of data consistent with t, that is to say $(w, u) \in S$ iff $t(w) = v$. The size of a sample is the sum of the length of the strings it is composed of: $|S| = \sum_{(w,u) \in S} |w| + |u|$.*

Definition 15. (Learning algorithm) *A $(\mathbb{T}, \mathbb{R})$-learning algorithm $\mathfrak{A}$ is a program that takes as input a sample for a function $t \in \mathbb{T}$ and outputs a representation from $\mathbb{R}$.*

The paradigm relies on the notion of a characteristic sample, adapted here for functions as in Chandlee et al. (2015).

Definition 16. (Characteristic sample) *For a $(\mathbb{T}, \mathbb{R})$-learning algorithm $\mathfrak{A}$, a sample CS is a characteristic sample of a function $t \in \mathbb{T}$ if for all samples $S \supseteq CS$, $\mathfrak{A}$ returns a representation r such that $\mathcal{L}(r) = t$.*

The learning paradigm can now be defined as follows.

Definition 17. (Identification in polynomial time and data) *A class $\mathbb{T}$ of functions is identifiable in polynomial time and data if there exists a $(\mathbb{T}, \mathbb{R})$-learning algorithm $\mathfrak{A}$ and two polynomial equations $p()$ and $q()$ such that:*

1. For any sample S of size m for $t \in \mathbb{T}$, $\mathfrak{A}$ returns a hypothesis $r \in \mathbb{R}$ in $\mathcal{O}(p(m))$ time.

2. For each representation $r \in \mathbb{R}$ of size n, with $t = \mathcal{L}(r)$, there exists a characteristic sample of t for $\mathfrak{A}$ of size at most $\mathcal{O}(q(n))$.

5.2 Learning when the tier is given

Prior to describing the approach we take to inducing the contents of a tier when $k = 2$, we note that learning any $OTSL_k$ function from positive data is relatively straightforward if the value of k and the tier Θ are known beforehand. In particular, although the OSLFIA presented in Chandlee et al. (2015) was designed only to learn OSL functions, it turns out that a minor modification allows us to extend their result to OTSL functions, so long as k and Θ are known beforehand. We summarize how this can be done below.

In its original form, the OSLFIA inevitably fails to learn any OTSL function that is not itself OSL (i.e., where $\Theta \neq \Delta$). Specifically, since the algorithm labels each landing state of a transition with the $k - 1$ suffix of its associated output, it will always incorrectly determine the landing state of one of more transitions when there is a long-distance dependency. Moreover, the exact way in which the resulting OSL transducer differs from the target OTSL transducer is somewhat unpredictable. As such, there does not seem to be a general approach for transforming the OSLFIA's output into an appropriate $OTSL_k$ transducer.

In cases where Θ is known beforehand, however, we can circumvent this issue by simply specifying that non-tier elements should be skipped over when labelling a state. In doing so, the algorithm will be able to find all of the necessary states as well as the correct landing state for each transition in the target $OTSL_k$ transducer. This modification of the OSLFIA is incorporated into the function `build_fst`, which is detailed in Algorithm 1. While this constitutes one important aspect of learning $OTSL_k$ functions, it is nonetheless a major challenge to determine the actual contents of Θ without *a priori* knowledge. Although inducing a k-tier for any value of k remains as an open problem, in the following section we describe how his can be done when $k = 2$.

5.3 Learning the contents of a 2-tier

Having shown that the OSLFIA can be modified to learn an $OTSL_k$ function f once Θ (the tier) is known, we now describe our approach to inducing Θ itself when $k = 2$. After this is done, Θ can simply be fed into the `build_fst` function in order to produce an $OTSL_2$ transducer that represents f.

The first step toward learning the contents of a 2-tier is to gain as much information as possible

Function `build_fst` $(S,\, \Theta,\, k)$:

  $C \leftarrow \{q_0, q_f\}$ with $q_0, q_f \notin \Theta^{\leq k-1}$;
  $s \leftarrow \text{lcp}(\{y | (x, y) \in S\})$;
  $q \leftarrow \text{Suff}_{\Theta}^{k-1}(s)$;
  $\text{Earliest}(q) \leftarrow \rtimes$;
  $\text{Out}(q) \leftarrow s$;
  $\delta \leftarrow \{(q_0, \rtimes, s, q)\}$;
  $R \leftarrow \{q\}$;
  while $R \neq \emptyset$ **do**
    $q \leftarrow \text{First}(R)$;
    $s \leftarrow \text{Earliest}(q)$;
    for *all $a \in \Sigma$ in alphabetical order* **do**
      if $\exists (w, u) \in S, x \in \Sigma^*$ *s.t.*
      $w = sax$ **then**
        $v \leftarrow \text{lcp}(\{y | \exists x, (sax, y) \in S\})$;
        $r \leftarrow \text{Suff}_{\Theta}^{k-1}(v)$;
        $\delta \leftarrow \delta \cup \{(q, a, \text{Out}(q)^{-1} \cdot v, r)\}$;
        if $r \notin R \cup C$ **then**
          $R \leftarrow R \cup \{r\}$;
          $\text{Earliest}(r) \leftarrow sa$;
          $\text{Out}(r) \leftarrow v$;
    if $\exists u$ *s.t.* $(s, u) \in S$ **then**
      $\delta \leftarrow \delta \cup \{(q, \ltimes, \text{Out}(q)^{-1} \cdot u, q_f)\}$
    $R \leftarrow R - \{q\}$;
    $C \leftarrow C \cup \{q\}$;
  return $\langle C, q_0, q_f, \Sigma, \Delta, \delta \rangle$

Algorithm 1: Building an $OTSL_k$ transducer when given Θ

about the prefix function f^p corresponding to f, based only on the evidence provided in the training sample. To do this, the function `estimate_fp`, shown in Algorithm 2, goes through every string x that is the prefix of at least one input string in the training data, and for every $a \in \Sigma$, it checks whether xa is also a prefix of some input string. If this is the case, there is enough information to determine $f^p(x)$. The function `estimate_fp` will then add the pair (x, z) to the set P, where z is the longest common prefix of $f(w)$ for all $(w, f(w)) \in S$ such that x is a prefix of w. We note that this z will be equal to $f^p(x)$ provided that the training data come from a subsequential function, and so this technique may be useful for learning other types of functions as well.

We further note, however, that by using this strategy, `estimate_fp` is only guaranteed to produce all the pairs $(x, f^p(x))$ necessary to discover the tier for total functions.

Function `estimate_fp`(S)**:**

> $P \leftarrow \emptyset$;
> $X \leftarrow \{x \mid x \in$ `Pref`$^*(w)$, where $(w, u) \in S\}$;
> $Y \leftarrow \{x \in X \mid (\forall a \in \Sigma)[xa \in X]\}$;
> **for** *each* $y \in Y$ **do**
> > $z \leftarrow$ `lcp`$(\{u \mid (w, u) \in S$, where $y \in$ `Pref`$^*(w)\})$;
> > $P \leftarrow P \cup \{(y, z)\}$
>
> **return** P;

Algorithm 2: Prefix function estimation

This is because, when there is no pair with the shape $(\rtimes pax \ltimes, f(pax))$ in the training data, `estimate_fp` does not know whether this is accidental (i.e., due to the finite nature of the training data) or because the function is undefined for all inputs of the shape $\rtimes pax \ltimes$. While the ability to accommodate partial functions would have practical applications for learning from natural language data, at present we leave the task of extending `estimate_fp` in this way to future research.

The full learning algorithm, which we call the OTSL$_2$ Function Inference Algorithm (OTSL2FIA) is shown in Algorithm 3. We assume that Σ and Δ are fixed and not part of the input to the learning problem (and that $k = 2$). Given a finite sample of training data, it first estimates the relevant prefix function with the set P, as described above, and begins with the hypothesis that $\Theta = \Delta$ (i.e., that all members of the output alphabet are on the target tier). Then, for each $a \in \Theta$, it looks through P for any evidence that a needs to be removed from Θ. To do this, it builds an auxiliary set `Match` that contains every $(p, f^p) \in P$ for which $\text{Suff}^1_\Theta(f^p(p)) = a$ under the current hypothesis for Θ. For each $x \in \Sigma \cup \{\ltimes\}$, it then checks whether $\text{cont}_f(x, p)$ is the same for all $(p, q) \in$ `Match`. If this is the case, a is added to the set `Keep`. However, if there is more than one value found for the contribution of some $x \in \Sigma \cup \{\ltimes\}$, it will instead remove a from Θ, since it cannot possibly be a member of the target 2-tier. If at any point some symbol gets removed from Θ, the set `Keep` is immediately emptied. This portion of the algorithm will run until every a in the current hypothesis for Θ gets added to the set `Keep`, in which case it knows it has found the canonical 2-tier of the target function.

Once the OTSL2FIA converges on the canon-

Data: Sample $S \subset \{\rtimes\}\Sigma^*\{\ltimes\} \times \Delta^*$
Result: An OTSL$_2$ transducer

$$\mathcal{T} = \langle C, q_0, q_f, \Sigma, \Delta, \delta \rangle$$

$P \leftarrow$ `estimate_fp`(S);
$\Theta \leftarrow \Delta$;
`Keep` $\leftarrow \emptyset$;
while *`Keep`* $\neq \Theta$ **do**
> **for** *each* $a \in \Theta$ **do**
> > `Match` $\leftarrow \{(p, q) \in P \mid$ `Suff`$^1_\Theta(q) = a\}$;
> > **for** *each* $\sigma \in \Sigma$ **do**
> > > $C_{\sigma, a} \leftarrow \{q^{-1} \cdot y \mid (p, q) \in$ `Match` $\land (p\sigma, y) \in P\}$;
> > > **if** $card(C_{\sigma, a}) > 1$ **then**
> > > > $\Theta \leftarrow \Theta - \{a\}$;
> > > > `Keep` $\leftarrow \emptyset$;
> >
> > $C_{\ltimes, a} \leftarrow \{q^{-1} \cdot y \mid (p, q) \in$ `Match` $\land (p, y) \in S\}$;
> > **if** $card(C_{\ltimes, a}) > 1$ **then**
> > > $\Theta \leftarrow \Theta - \{a\}$;
> > > `Keep` $\leftarrow \emptyset$;
> >
> > **if** $a \in \Theta$ **then**
> > > `Keep` $\leftarrow$ `Keep` $\cup \{a\}$

$\mathcal{T} \leftarrow$ `build_fst`$(S, \Theta, 2)$;
return $\mathcal{T}$

Algorithm 3: OTSL2FIA

ical 2-tier Θ, the final step is simply to feed Θ and the sample S into the function `build_fst` shown above in Algorithm 1 (further specifying that $k = 2$). Under the assumption that the training sample contains the appropriate evidence, as described in the following section, this will produce an OTSL$_2$ transducer which represents the target OTSL$_2$ function.

5.4 Theoretical results

Here we establish several theoretical results, which culminate in the theorem that the OTSL2FIA identifies the class of total OTSL$_2$ functions in polynomial time and data.

In what follows, we let f be the target OTSL$_2$ function, $\Theta_\diamond$ be its canonical 2-tier, and $\mathcal{T}^\diamond = \langle Q_\diamond, q_0, q_f, \Sigma, \Delta, \delta_\diamond \rangle$ be its target transducer as defined by Theorem 2. We furthermore let Θ be the OTSL2FIA's final tier hypothesis, and $\mathcal{T} = \langle Q, q_0, q_f, \Sigma, \Delta, \delta \rangle$ be the transducer that is constructed on the input.

Lemma 6. (Polynomial time) *For any input sample S, the OTSL2FIA produces $\mathcal{T}$ in time polynomial in the size of S.*

Proof. Let $n = \sum_{(w,u)\in S}|w|$, $m = max\{|u| : (w,u) \in S\}$, $p = max\{|w| : (w,u) \in S\}$, and $s = \texttt{card}(S)$. We note that these are all linear in the size of the sample.

The OTSL2FIA starts by calling $\texttt{estimate\_fp}$. This function first determines all of the input prefixes present in S, which takes n steps. Then $\texttt{estimate\_fp}$ checks, for each prefix x and all $a \in \Sigma$, whether xa is also an input prefix in S. There are at most sm prefixes in S, so this takes at most $\texttt{card}(\Sigma) \cdot (sm)n$ steps. Finally, for a subset of the input prefixes, $\texttt{estimate\_fp}$ determines $\texttt{lcp}(\{u \mid (w,u) \text{ s.t. } x \in \texttt{Pref}^*(w)\})$, which with an appropriate data structure (for instance a prefix tree) can be done in nm steps. The overall computation time of $\texttt{estimate\_fp}$ is thus $\mathcal{O}(n + (sm)n + (sm)(nm))$, which is quartic in the size of the learning sample.

The portion of the OTSL2FIA that determines the tier is now run. After i elements have been removed from Θ, the combined *while/for* loop can run up to $|\Delta| - i$ times, and can only remove up to $|\Delta|$ items, so the loop will be used fewer than $|\Delta|^2$ times, which is a constant. This main loop first gathers all $(w,u) \in P$ that meet a certain criterion into the set $\texttt{Match}$, which can be done in $\texttt{card}(P)m = (sm)m = sm^2$ steps. Next, the main loop enters a *for* loop that is used $\texttt{card}(\Sigma)$ times (a constant) and which attempts to calculate the contribution of $\sigma \in \Sigma$ using each $(w,u) \in \texttt{Match}$ if it can find $(w\sigma, v) \in P$. We note that $\texttt{card}(\texttt{Match})$ will be at most sm, that finding $(w\sigma, v) \in P$ takes at most smp steps, and that calculating the contribution takes at most m steps. The main loop then attempts to calculate the contribution of $\ltimes$ using each $(w,u) \in \texttt{Match}$ if it can find $(w,v) \in S$. We note that finding $(w,v) \in S$ takes at most n steps, and that calculating the contribution takes at most m steps. The overall computation time of this portion of the algorithm is thus $\mathcal{O}(sm^2 + sm(smp+m) + sm(n+m))$, which is quintic in the size of the learning sample.

Finally, the OTSL2FIA feeds Θ and S to the function $\texttt{build\_fst}$. As noted above, this fuction incorporates a simple modification to the state-labelling process in Chandlee et al.'s (2015) OSLFIA. While this change allows it to build an OTSL transducer once the tier is known, it does not affect computation time. This final step of the OTSL2FIA therefore runs in time quadratic

in the size of the learning sample (for OSLFIA time complexity proofs, see Chandlee et al., 2015). Since each portion of the OTSL2FIA runs in time polynomial in the size of the sample, with the highest complexity being quintic, the overall computation time of the algorithm is therefore polynomial in the size of the learning sample.

$\square$

The remaining lemmas of this section will show that for each total OTSL_2 function f, there is a finite kernel of data consistent with f that is a characteristic sample for OTSL2FIA, which we call an OTSL2FIA seed.

Definition 18. (Seed) *Given $\mathcal{T}^\diamond$, a sample S contains a seed if:*

1. *For all $q \in Q_\diamond$, $(\rtimes w_q \ltimes, f(w_q)) \in S$.*

2. *For all $(q, a, u, q') \in \delta_\diamond$ such that $q' \neq q_f$ and $a \in \{\rtimes\} \cup \Sigma$, and for all pairs $b, c \in \Sigma$:*

   (a) *$(\rtimes w_q a\ltimes, f(w_q a)) \in S$*

   (b) *$(\rtimes w_q ab\ltimes, f(w_q ab)) \in S$*

   (c) *$(\rtimes w_q abcx\ltimes, f(w_q abcx)) \in S$, where $x \in \Sigma^*$*

Lemma 7. *If a learning sample S contains a seed, then the OTSL2FIA can determine $\text{cont}_f(x, w)$ for all $w \in \Sigma^*$ and all $x \in \Sigma \cup \{\ltimes\}$.*

Proof. Let us start with some string $\rtimes wby\ltimes$ such that $b \in \Sigma$ and $w, y \in \Sigma^*$. Since $\mathcal{T}^\diamond$ is OTSL_2, it will be in the state corresponding to $\texttt{Suff}^1_{\Theta_\diamond}(f^p(w))$ immediately prior to reading b when processing $\rtimes wby\ltimes$. Let us call this state q'. The most recent transition that $\mathcal{T}^\diamond$ will have traversed is (q, a, u, q'). The target function is OTSL_2, and so it is the case that either $w = w_q a$ or else can be replaced thereby since $\texttt{Suff}^1_{\Theta_\diamond}(f^p(w)) = \texttt{Suff}^1_{\Theta_\diamond}(f^p(w_q a))$ and therefore $\text{cont}_f(b, w) = \text{cont}_f(b, w_q a)$.

Now let us start with some string $\rtimes w\ltimes$ such that $w \in \Sigma^*$. When $\mathcal{T}^\diamond$ reads $\rtimes w\ltimes$, it will be in the state corresponding to $\texttt{Suff}^1_{\Theta_\diamond}(f^p(w))$ immediately prior to reading $\ltimes$. Let us call this state q'. The most recent transition that $\mathcal{T}^\diamond$ will have traversed is (q, a, u, q'). The target function is OTSL_2, and so it is the case that either $w = w_q a$ or else can be replaced thereby since $\texttt{Suff}^1_{\Theta_\diamond}(f^p(w)) = \texttt{Suff}^1_{\Theta_\diamond}(f^p(w_q a))$ and therefore $\text{cont}_f(\ltimes, w) = \text{cont}_f(\ltimes, w_q a)$.

Now recall that $f^p(w) = \texttt{lcp}(\{u \mid u = f(wx) \wedge x \in \Sigma^*\})$. We do not actually need the entirety of this infinite set to determine $f^p(w)$, it is sufficient to use a set containing $f(w)$ and one $f(wax)$ for each $a \in \Sigma$ where $x \in \Sigma^*$ because every $x \in \Sigma^*$ is either λ or begins with some $a \in \Sigma$. Let us call such a set a *support* for determining $f^p(w)$. The function `estimate_fp` takes every prefix p present in S and checks whether a support for determining $f^p(p)$ exists in S. Then, if a support exists, `estimate_fp` adds (p, q) to the set P, where $q = \mathtt{lcp}(\{u \mid (w, u) \in S \wedge p \in \mathtt{Pref}^*\})$, that is $q = f^p(p)$.

By the definition of the seed, for every transition $(q, a, u, q') \in \delta_\diamond$ such that $q' \neq q_f$, the learner will see $\ltimes w_q a \rtimes$, $\ltimes w_q ab \rtimes$ for all $b \in \Sigma$, and at least one input string $\ltimes w_q abcx \rtimes$ for all pairs $b, c \in \Sigma$. We therefore know that for any pair of input strings $\ltimes w \rtimes$ and $\ltimes wby \rtimes$ in the domain of f such that $b \in \Sigma$ and $x \in \Sigma^*$, the seed will contain all the input strings necessary to build supports for determining $f^p(w_q a)$ and $f^p(w_q ab)$ such that $\mathtt{tails}_f(w_q a) = \mathtt{tails}_f(w)$.

By Remark 1, we know that $\mathtt{cont}_f(b, w) = f^p(w)^{-1} \cdot f^p(wb)$ for all $b \in \Sigma$ and $w \in \Sigma^*$. It is therefore the case that for every $\ltimes wby \rtimes$, the algorithm can determine $\mathtt{cont}_f(b, w_q a) = f^p(w_q a)^{-1} \cdot f^p(w_q ab) = \mathtt{cont}_f(b, w)$. Also by Remark 1, we know that $\mathtt{cont}_f(\ltimes, w) = f^p(w)^{-1} \cdot f(w)$ for all $w \in \Sigma^*$. It is therefore the case that for every $\ltimes w \rtimes$, the algorithm can determine $\mathtt{cont}_f(\ltimes, w_q a) = f^p(w_q a)^{-1} \cdot f(w_q a) = \mathtt{cont}_f(\ltimes, w)$. $\square$

Lemma 8. (Tier convergence) *If a learning sample S contains a seed, then $\Theta = \Theta_\diamond$.*

Proof. The OTSL2FIA starts with $\Theta = \Delta$, and so either $\Theta = \Theta_\diamond$ already, or else $\Theta \supset \Theta_\diamond$.

We know from Lemma 5 that if $\Theta \supset \Theta_\diamond$, there will exist a pair of input strings w_1 and w_2 in the domain of f such that $\mathtt{Suff}^1_\Theta(f^p(w_1)) = \mathtt{Suff}^1_\Theta(f^p(w_2)) = a$ for some $a \in (\Theta - \Theta_\diamond)$ and $\mathtt{cont}_f(x, w_1) \neq \mathtt{cont}_f(x, w_2)$ for some $x \in \Sigma \cup \{\ltimes\}$. We know from Lemma 7 that every $\ltimes w \rtimes$ in the domain of f has at least one corresponding $(w', f^p(w')) \in P$ and at least one corresponding $(w'a, f^p(w'a)) \in P$ for each $a \in \Sigma$, where $\mathtt{Suff}^1_\Theta(f^p(w)) = \mathtt{Suff}^1_\Theta(f^p(w'))$ and so $\mathtt{cont}_f(x, w) = \mathtt{cont}_f(x, w')$ for all $x \in \Sigma \cup \{\ltimes\}$. The algorithm will thus be able to calculate and check all the relevant contributions necessary to flag and remove at least one $a \in (\Theta - \Theta^\diamond)$ when $\Theta \supset \Theta^\diamond$.

Conversely, there will be no pair of input strings w_3 and w_4 in the domain of f such that $\mathtt{cont}_f(x, w_3) \neq \mathtt{cont}_f(x, w_4)$ for some $x \in \Sigma \cup \{\ltimes\}$ when $\mathtt{Suff}^1_\Theta(f^p(w_3)) = \mathtt{Suff}^1_\Theta(f^p(w_3)) = c$ for some $c \in \Theta^\diamond$. When $\Theta = \Theta_\diamond$, then, the algorithm will add all $a \in \Theta$ to `Keep` and pass `Keep` $= \Theta = \Theta_\diamond$ to the `build_fst` function. $\square$

Lemma 9. (Transducer convergence) *If a learning sample S contains a seed then $(q_0, \ltimes w, u, r) \in \delta^* \Longleftrightarrow (q_0, \ltimes w, u, r) \in \delta_\diamond^*$.*

Lemma 10. (Characteristic Sample) *Any learning sample containing a seed is a characteristic sample for the OTSL2FIA.*

We do not include the proofs of Lemmas 9 and 10 here, as they are a trivial extension of analogous proofs in Chandlee et al. (2015, Lemmas 7 and 8). Again, the generalization requires only that each instance of $\mathtt{Suff}^{k-1}$ be replaced by $\mathtt{Suff}^{k-1}_\Theta$ in order for the proofs hold for any OTSL$_k$ transducer, provided that the target tier is passed to `build_fst`. We further note that when the target transducer is one that computes a total OTSL2 function with its canonical tier, a seed (as defined in Definition 18 above) is a superset of that required by Chandlee et al. (2015, Definition 11).

Lemma 11. (Polynomial data) *Given an OTSL2 transducer $\mathcal{T}^\diamond$, there exists a seed for the OTSL2FIA that is of size polynomial in the size of $\mathcal{T}^\diamond$.*

Proof. Let $\mathcal{T}^\diamond = \langle Q_\diamond, q_0, q_f, \Sigma, \Delta, \delta_\diamond \rangle$. For item 1 in Definition 18 there are $\mathtt{card}(Q_\diamond)$ corresponding input-output pairs $(w_q, f(w_q))$ in a seed. For each of these pairs, it is the case that $|\ltimes w_q \rtimes| \leq \mathtt{card}(Q_\diamond)$ and it is the case that $|f(w_q)| \leq \sum_{(q,a,u,q') \in \delta_\diamond} |u|$. We denote the latter quantity with $x_\diamond = \sum_{(q,a,u,q') \in \delta_\diamond} |u|$ and note that $x_\diamond = \mathcal{O}(|\mathcal{T}^\diamond|)$. The overall length of the inputs in the portion of the seed contributed by item 1 is thus in $\mathcal{O}(\mathtt{card}(Q_\diamond)^2)$. The overall length of the outputs in the portion of the seed contributed by item 1 is thus in $\mathcal{O}(\mathtt{card}(Q_\diamond) \cdot x_\diamond)$. We note that both of these are quadratic in the size of $\mathcal{T}^\diamond$.

For items 2a, 2b, and 2c in Definition 18, there are respectively 1, $\mathtt{card}(\Sigma)$, and $\mathtt{card}(\Sigma)^2$ corresponding input-output pairs per transition $(q, a, u, q') \in \delta_\diamond$. Factoring out the constant $\mathtt{card}(\Sigma)$ gives us $3 \cdot \mathtt{card}(\delta_\diamond)$ pairs. For the pairs contributed by item 2c, we can restrict ourselves to pairs $(\ltimes w_q abc \rtimes, f(w_q abc))$, since f is a total function. For each pair, we have $|\rtimes w \ltimes| \leq$ $\mathrm{card}(Q_\diamond) + 3$ and $|f(w)| \leq \sum_{(q,a,u,q')\in\delta_\diamond} |u| + 3m$, where $m = max\{|u| : (q,a,u,q') \in \delta_\diamond\}$. With this last quantity denoted $y_\diamond$, we note that $y_\diamond = \mathcal{O}(|\mathcal{T}^\diamond|)$. The overall length of the inputs in the portion of the seed contributed by item 2 is therefore in $\mathcal{O}((3 \cdot \mathrm{card}(\delta_\diamond))(\mathrm{card}(Q_\diamond) + 3) = \mathcal{O}(\mathrm{card}(\delta_\diamond) \cdot \mathrm{card}(Q_\diamond) + \mathrm{card}(\delta_\diamond))$, and the overall length of the outputs in the portion of the seed contributed by item 2 is in $\mathcal{O}(3 \cdot \mathrm{card}(\delta_\diamond) \cdot y_\diamond)$. Both of these are quadratic in the size of $\mathcal{T}^\diamond$.

Altogether, then, the size of the seed is quadratic in the size of the target transducer. $\qquad \square$

Theorem 3. *The OTSL2FIA identifies the OTSL$_2$ functions in polynomial time and data.*

Proof. Immediate from Lemmas 6, 8, 9, 10, and 11. $\qquad \square$

6 Discussion

The OTSL functions introduced in this paper are capable of modelling many of the attested long-distance phonological processes. These processes can be assimilatory like sibilant harmony in Samala (see Section 1), but can also be dissimilatory. For example, Georgian exhibits a pattern of liquid dissimilation, in which /r/ surfaces as [l] when preceded at any distance by another [r] (e.g., /aprik'-uri/ → [aprik'uli] 'African'; Odden, 1994). Interestingly, the dissimilation does not occur if there is an intervening [l] (e.g., /kartl-uri/ → [kartluri] 'Kartvelian'). The OTSL functions are fully capable of representing such *blocking* effects, as shown in Figure 2. To avoid cluttering the figure, we omit the final state and all of its incoming transitions (which would be labelled $\ltimes:\lambda$).

It is worth pointing out that the processes in Samala and Georgian apply in opposite directions. In Samala, the trigger is the *rightmost* sibilant, whereas in Georgian it is the *leftmost* liquid. This distinction can be captured by assuming that input strings are read from left-to-right in the Georgian case (i.e., the process is progressive), but from right-to-left in the Samala case (i.e., the process is regressive). The direction of reading, then, divides the OTSL functions into two overlapping but distinct classes which we call L-OTSL (which read from the left) and R-OTSL (which read from the right), following Heinz and Lai (2013) and Chandlee et al. (2015) who make the same distinction

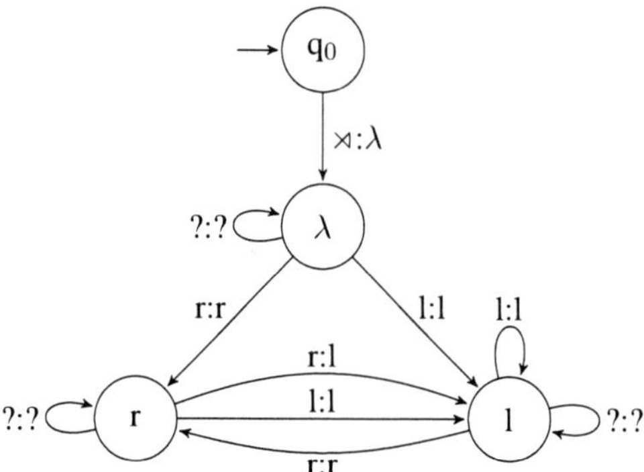

Figure 2: An OTSL$_2$ transducer that models unbounded liquid dissimilation with blocking, where ? represents any symbol that is not [l] or [r].

for the subsequential and OSL functions, respectively.

As mentioned above, the OTSL2FIA outlined in Section 5 only succeeds in learning total functions and is designed specifically to learn OTSL$_2$ functions. The algorithm exploits the fact that the largest possible 2-tier for an OTSL$_2$ function f is a superset of every other 2-tier for f, and will accordingly never run the risk of removing an element that would need to be subsequently re-added to the tier. However, it is not clear that this strategy will succeed for higher values k, which may be needed to model certain types of patterns. For example, a reviewer raises the complex case of retroflexion harmony targeting /n/ in Sanskrit (also known as *nati*) as one such pattern. A formal analysis provided by Graf and Mayer (2018) uses a class of stringsets that they call Input-Output Tier-based Strictly Local (IO-TSL). IO-TSL formal languages are like TSL languages except that input symbols are projected onto the tier based on (i) the surrounding context of input symbols and (ii) the symbols that precede it on the tier that has been projected so far. Under this analysis, Sanskrit n-retroflexion requires $k = 3$ on the projected tier.

Finally, while the OTSL class can model long-distance processes, it can only do so when no more than a single tier is required. That is, a language that simultaneously exhibits patterns of, e.g., sibilant harmony and liquid dissimilation would not be OTSL for any value of k. Further exploration of these issues will allow us to better understand the computational properties of phonological transformations and to establish a boundary of complexity that is both necessary and sufficient for capturing the full range of possible phonological systems.

7 Conclusion

This paper has provided both a language-theoretic and an automata-theoretic characterization of the OTSL class of functions, which is relevant for modelling long-distance phonological processes as string-to-string transformations. We further demonstrated that by generalizing previous research on OSL functions to the OTSL class, any $OTSL_k$ function can be learned once the tier is known. Finally, we introduced an algorithm for efficiently learning any total $OTSL_2$ function from positive data, even when a relevant tier is not given *a priori*. To our knowledge, this is the first algorithm to accomplish this for input-output mappings rather than phonotactics. In future research, we aim to extend this result in multiple ways: to partial functions, to any value of k, and to processes requiring multiple tiers.

Acknowledgements

Special thanks to Jane Chandlee, participants of the Third Subregular Workshop at Stony Brook University, and three anonymous reviewers. This research was supported by the Social Sciences and Humanities Research Council of Canada.

References

Richard B. Applegate. 1972. *Ineseño Chumash grammar*. Doctoral dissertation, University of California, Berkeley.

Jane Chandlee. 2014. *Strictly Local phonological processes*. Ph.D. thesis, University of Delaware.

Jane Chandlee, Rémi Eyraud, and Jeffrey Heinz. 2014. Learning strictly local subsequential functions. *Transactions of the Association for Computational Linguistics*, 2:491–503.

Jane Chandlee, Rémi Eyraud, and Jeffrey Heinz. 2015. Output strictly local functions. In *Proceedings of the 14th Meeting on the Mathematics of Language*.

George N. Clements. 1980. *Vowel harmony in nonlinear generative phonology: an autosegmental model*. Indiana University Linguistics Club, Bloomington, IN.

Colin de la Higuera. 1997. Characteristic sets for polynomial grammatical inference. *Machine Learning*, 27(2):125–138.

E. Mark Gold. 1967. Language identification in the limit. *Information and Control*, 10:447–474.

John A. Goldsmith. 1990. *Autosegmental and metrical phonology*. Blackwell, Oxford.

Maria Gouskova and Gillian Gallagher. 2019. Inducing nonlocal constraints from baseline phonotactics. *Natural Language and Linguistic Theory*.

Thomas Graf and Connor Mayer. 2018. Sanskrit n-retroflexion is input-output tier-based strictly local. In *Proceedings of SIGMORPHON 2018*, pages 151–160.

Gunnar Ólafur Hansson. 2010. *Consonant harmony: long-distance interaction in phonology*. University of California Press, Berkeley, CA.

Bruce Hayes and Colin Wilson. 2008. A maximum entropy model of phonotactics and phonotactic learning. *Linguistic Inquiry*, 39:379–440.

Jeffrey Heinz and Regine Lai. 2013. Vowel harmony and subsequentiality. In *Proceedings of the 13th Meeting on the Mathematics of Language*, pages 52–63, Sofia, Bulgaria.

Jeffrey Heinz, Chetan Rawal, and Herbert G. Tanner. 2011. Tier-based strictly local constraints for phonology. In *Proceedings of the 49th Annual Meeting of the Association for Computational Linguistics*, pages 58–64, Portland, OR. Association for Computational Linguistics.

Adam Jardine, Jane Chandlee, Rémi Eyraud, and Jeffrey Heinz. 2014. Very efficient learning of structured classes of subsequential functions from positive data. In *Proceedings of the Twelfth International Conference on Grammatical Inference*, volume 34, pages 94–108.

Adam Jardine and Jeffrey Heinz. 2016. Learning tier-based strictly 2-local languages. *Transactions of the Association for Computational Linguistics*, 4:87–98.

Adam Jardine and Kevin McMullin. 2017. Efficient learning of Tier-based Strictly k-Local languages. In *Proceedings of Language and Automata Theory and Applications, 11th International Conference*, Lecture Notes in Computer Science. Springer.

Aaron Kaplan. 2008. *Noniterativity is an emergent property of grammar*. Ph.D. thesis, University of California Santa Cruz.

Kevin McMullin. 2016. *Tier-based locality in long-distance phonotactics: learnability and typology*. Ph.D. thesis, University of British Columbia.

David Odden. 1994. Adjacency parameters in phonology. *Language*, 70:289–330.

José Oncina and Pedro García. 1991. Inductive learning of subsequential functions. Techical Report DSIC II-34, University Politecnia de Valencia.

José Oncina, Pedro García, and Enrique Vidal. 1993. Learning subsequential transducers for pattern recognition tasks. *IEEE Transactions on Pattern Analysis and Machine Intelligence*, 15:448–458.

James Rogers and Geoffrey K. Pullum. 2011. Aural pattern recognition experiments and the subregular hierarchy. *Journal of Logic, Language and Information*, 20(3):329–342.

Learning with Partially Ordered Representations

Jane Chandlee
Tri-Co Department of Linguistics
Haverford College
jchandlee@haverford.edu

Rémi Eyraud
QARMA team, LIS
Aix-Marseille University
remi.eyraud@lis-lab.fr

Jeffrey Heinz
Department of Linguistics
Institute for Advanced Computational Science
Stony Brook University
jeffrey.heinz@stonybrook.edu

Adam Jardine
Department of Linguistics
Rutgers University
adam.jardine@rutgers.edu

Jonathan Rawski
Department of Linguistics
Institute for Advanced Computational Science
Stony Brook University
jonathan.rawski@stonybrook.edu

Abstract

This paper examines the characterization and learning of grammars defined with enriched representational models. Model-theoretic approaches to formal language theory traditionally assume that each position in a string belongs to exactly one unary relation. We consider unconventional string models where positions can have multiple, shared properties, which are arguably useful in many applications. We show the structures given by these models are partially ordered, and present a learning algorithm that exploits this ordering relation to effectively prune the hypothesis space. We prove this learning algorithm, which takes positive examples as input, finds the most general grammar which covers the data.

1 Introduction

Foundational connections between formal languages, finite-state automata, and logic have been known for decades (Büchi, 1960; Thomas, 1997). Logical approaches are advantageous since they flexibly admit different representations. In many domains, such as biological sequencing or linguistics, shared properties of symbols in sequences provide information currently ignored by string-based inference algorithms, which largely focus on learning automata (de la Higuera, 2010). Here we explore the idea that domain-specific knowledge can be encoded representationally via model theory (Libkin, 2004), and shows how these representations can facilitate pattern learning.

This paper synthesizes results in grammatical inference and model theory to present a novel algorithm which learns classes of formal languages using enriched representations of strings. In fact, our model-theoretic approach immediately generalizes these results to arbitrary data structures. Here we are concerned with the learning of those formal languages which can be defined via a set of structural constraints, such as the Strictly k-Local and Strictly k-Piecewise languages (Rogers and Pullum, 2011; Rogers et al., 2010). Models of strings in the languages must not contain these forbidden structures (Rogers et al., 2013). Specifically, we define a learner whose hypothesis space is structured as a partial order by the relational signature of the particular model theory. We show how to traverse this space bottom-up from positive data to find a grammar which covers the data with the most general constraints.

The paper is structured as follows: Section 2 provides mathematical preliminaries in model theory. Section 3 characterizes ordering relations over these structures. Section 4 generalizes the grammars employed in string extension and lattice-based learning (Heinz, 2010; Heinz et al., 2012) to show how these model theoretic structures can define classes of formal languages. Section 5 discusses some entailments our learning al-

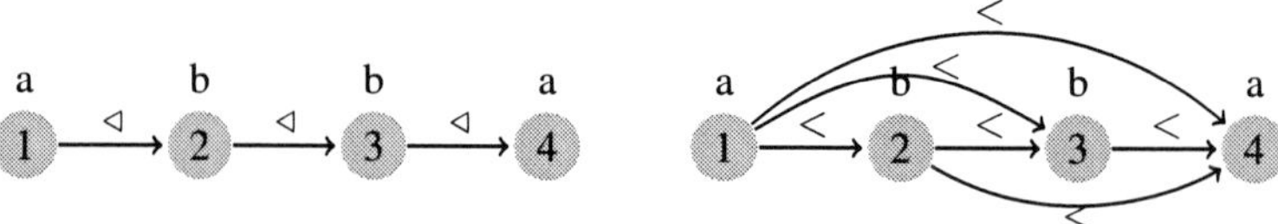

Figure 1: Visualizations of the successor (left) and precedence (right) models of *abba*.

gorithm takes advantage of. Section 6 defines a learning problem and criteria for selecting adequate solutions. Section 7 presents a general-to-specific, bottom-up algorithm which provably satisfies the learning criteria. Section 8 concludes the paper.

2 Preliminaries

2.1 Elements of Language Theory

The set of all possible finite strings of symbols from a finite alphabet Σ and the set of strings of length $\leq n$ are Σ^* and $\Sigma^{\leq n}$, respectively. The unique empty string is represented with λ. The length of a string w is $|w|$, so $|\lambda| = 0$. If u and v are two strings then we denote their concatenation with uv. If w is a string and σ is the ith symbol in w then $w_i = \sigma$, so $abcd_2 = b$.

The set of prefixes of w, $\mathrm{Pref}(w)$, is $\{p \in \Sigma^* \mid (\exists s \in \Sigma^*)[w = ps]\}$, the set of suffixes of w, $\mathrm{Suff}(w)$, is $\{s \in \Sigma^* \mid (\exists p \in \Sigma^*)[w = ps]\}$, the set of substrings, $\mathrm{Substr}(w)$, is $\{u \in \Sigma^* \mid (\exists l, r \in \Sigma^*)[w = lur]\}$, and the set of subsequences, $\mathrm{Subseq}(w) = u_1 u_2 \cdots u_n | \exists v_0 \cdot v_1 \cdots v_n \in \Sigma^*[w = v_0 u_1 v_1 \cdots u_n v_n]$

2.2 Elements of Finite Model Theory

Model theory, combined with logic, provides a powerful way to study and understand mathematical objects with structures (Enderton, 2001). In this paper we only consider finite relational models (Libkin, 2004) of strings in Σ^*.

Definition 1 (Models). *A model signature is a tuple $S = \langle D; R_1, R_2, \ldots, R_m \rangle$ where the domain D is a finite set, and each R_i is a n_i-ary relation over the domain. A model for a set of objects Ω is a total, one-to-one function from Ω to structures whose type is given by a model signature.*

For example, a conventional model for strings in Σ^* is given by the signature $\Gamma^\triangleleft \overset{def}{=} \langle D; \triangleleft, [R_\sigma]_{\sigma \in \Sigma} \rangle$ and the function $M^\triangleleft : \Sigma^* \to \Gamma^\triangleleft$ such that $M^\triangleleft(w) \overset{def}{=} \langle D^w; \triangleleft, [R_\sigma^w]_{\sigma \in \Sigma} \rangle$ where $D^w \overset{def}{=} \{1, \ldots, |w|\}$ is the domain,

$\triangleleft \overset{def}{=} \{(i, i+1) \in D \times D\}$ is the successor relation which orders the elements of the domain, and $[R_\sigma^w]_{\sigma \in \Sigma}$ is a set of $|\Sigma|$ unary relations such that for each $\sigma \in \Sigma$, $R_\sigma^w \overset{def}{=} \{i \in D^w \mid w_i = \sigma\}$. We will usually omit the superscript w since it will be clear from the context.

For example, with $\Sigma = \{a, b, c\}$ and the model above for strings, we have $M^\triangleleft(abba) = \langle D = \{1, 2, 3, 4\}; \triangleleft = \{(1, 2), (2, 3), (3, 4)\}, R_a = \{1, 4\}, R_b = \{2, 3\}, R_c = \emptyset \rangle$.

Figure 1 illustrates $M^\triangleleft(abba)$ on the left.

Another conventional model is the precedence model, with the signature $\Gamma^< \overset{def}{=} \langle D; <, [R_\sigma]_{\sigma \in \Sigma} \rangle$. It differs from the successor model only in that the order relation is defined with general precedence $< \overset{def}{=} \{(i, j) \in D \times D \mid i < j\}$ (Büchi, 1960; McNaughton and Papert, 1971; Rogers et al., 2013). Under this signature, the string *abba* has the following model.

$M^<(abba) = \langle D = \{1, 2, 3, 4\}; <= \{(1, 2), (1, 3), (1, 4), (2, 3), (2, 4), (3, 4)\}, R_a = \{1, 4\}, R_b = \{2, 3\}, R_c = \emptyset \rangle$.

Figure 1 illustrates $M^<(abba)$ on the right.

The model-theoretic framework is not unique to strings. It extends to arbitrary data structures by expanding parts of the model signature. For example, Rogers (2003) describes a model-theoretic characterization of trees of arbitrary dimensionality where the domain D is a Gorn tree domain (Gorn, 1967). This is a hereditarily prefix closed set D of node addresses, that is to say, for every $d \in D$ with $d = \alpha i$, where $i \in \mathbb{N}$, $\alpha \in \mathbb{N}^*$ it holds that $\alpha \in D$, and for every $d \in D$ with $d = \alpha i \neq \alpha 0$, then $\alpha(i - 1) \in D$.

In this view, a string may be called a one-dimensional or unary-branching tree, since it has one axis along which its nodes are ordered. In a standard tree mdoel signature, the set of nodes is ordered by two binary relations, "dominance" and "immediate left-of". Suppose s is the mother of two nodes t and u in some standard tree, and also assume that t precedes u. Then we might say that s dominates the string tu. Standard or two-dimensional trees, then, relate nodes to one-dimensional trees (strings) by immediate dominance. A three-dimensional tree relates nodes to two-dimensional, i.e. standard trees, corresponding to Tree-Adjoining Grammar derivations. In general, a d-dimensional tree is a set of nodes ordered by d dominance relations such that the n-th dominance relation relates nodes to $(n-1)$-dimensional trees (for $d = 1$, single nodes are zero-dimensional trees).

While a Gorn tree domain as written encodes these dominance and precedence relations implicitly, we may explicitly write them out model-theoretically so that a signature for a Σ-labeled 2-d tree is $\Gamma^{\lhd\prec} = \langle D; \lhd, \prec, [R_\sigma]_{\sigma \in \Sigma} \rangle$ where $\lhd$ is the "immediate dominance" relation and $\prec$ is the "immediate left-of" relation. Model signatures that include transitive closure relations of each of these have also been studied.

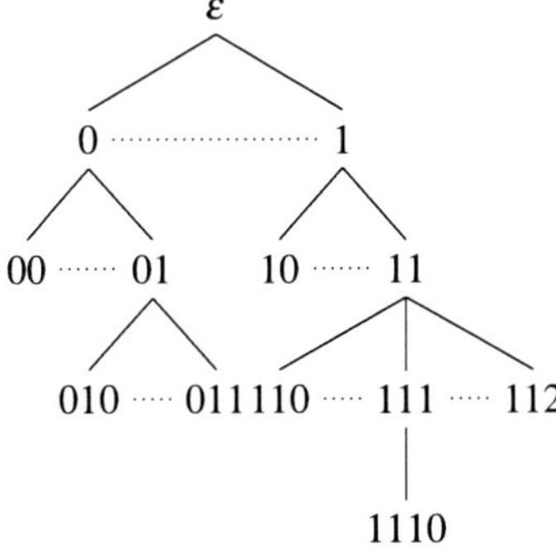

Figure 2: 2-dimensional tree model. Dominance and precedence relations shown with solid/dashed and dotted lines, respectively

2.3 Unconventional Word Models

Whereas Rogers (2003) generalized conventional word models to trees, here we generalize word models in a different way. Conventional string models are the successor and precedence models introduced previously. What makes these models conventional is the unary relations which essentially label each domain element with a single, mutually exclusive, property: the property of being some $\sigma \in \Sigma$.

In contrast, unconventional models for strings recognize that distinct alphabetic symbols may share properties, and expands the model signature by including these properties as unary relations (Strother-Garcia et al., 2016; Vu et al., 2018). For example, a conventional model of

$\Sigma = \{\mathtt{a}, \dots, \mathtt{z}, \mathtt{A}, \dots, \mathtt{Z}\}$ would include 52 unary relations, one for each lowercase and capital letter. On the other hand, an unconventional model might only include 27: 26 for the letters, and one unary relation $\mathtt{Capital}$. Then, letters $\mathtt{A}$ and $\mathtt{a}$ share the 'A' property and $\mathtt{A}$ additionally has the property of being $\mathtt{Capital}$.

In linguistics, speech sounds are commonly decomposed into binary features based on their phonetic properties. So the set of segments $\{\mathtt{z},\mathtt{ʒ},\mathtt{d},\mathtt{b},\mathtt{g}, \dots\}$ all share the property $\mathtt{+Voice}$, meaning the vocal cords are activated, while the segments $\{\mathtt{s},\mathtt{ʃ},\mathtt{t},\mathtt{p},\mathtt{k}, \dots\}$ share the property $\mathtt{-Voice}$, meaning the vocal cords are not activated. Thus unconventional models may refer to individual features in defining grammatical constraints, rather than each individual segment.

Different representations of strings and trees provide a unified perspective on well-known subclasses of the regular languages from a model-theoretic and logical perspective (Thomas, 1997; Rogers et al., 2013). However, they also open up new doors for grammatical inference by allowing one to consider other models for strings (Strother-Garcia et al., 2016; Vu et al., 2018).

3 Subfactors, Superfactors, Ideals and Filters

We sometimes refer to the model of a string w as a *structure*. However, structures are more general in that they correspond to any mathematical structure conforming to the model signature. As such, while a model of a string w will always be a structure, a structure will not always be a model of a string w. The *size* of a structure S, denoted $|S|$, coincides with the cardinality of its domain.

We next wish to introduce a partial ordering over structures. To do so, we must define the terms *connected, restriction*, and *factor*. For each structure $S = \langle D; \lhd, R_1, \dots R_n \rangle$ let the binary "connectedness" relation C be defined as follows.

$$C \stackrel{def}{=} \{(x,y) \in D \times D \mid \exists i \in \{1 \dots n\}, \exists (x_1 \dots x_m) \in R_i, \exists s, t \in \{1 \dots m\}, x = x_s, y = x_t\}$$

Informally, domain elements x and y belong to C provided they belong to some non-unary relation. Let C^* denote the symmetric transitive closure of C.

Definition 2 (Connected structure). *A structure* $S = \langle D; \lhd, R_1, R_2, \dots, R_n \rangle$ *is* connected *iff for all* $x, y \in D$, $(x,y) \in C^*$.

For example, $M^{\lhd}(abba)$ above is a connected structure. However, the structure $S_{ab.\,ba}$ shown below which is identical to $M^{\lhd}(abba)$ except it omits the pair $(2,3)$ from the order relation is not connected since none of $(1,3),(1,4)$, $(2,3)$ nor $(2,4)$ belong to C^*. $S_{ab,\,ba} = \langle D = \{1,2,3,4\}; \lhd = \{(1,2),(3,4)\}, R_a = \{1,4\}, R_b = \{2,3\}, R_c = \emptyset\rangle$

<table>
<tr><td>a</td><td></td><td>b</td><td>b</td><td></td><td>a</td></tr>
<tr><td>(1)</td><td>—▷→</td><td>(2)</td><td>(3)</td><td>—▷→</td><td>(4)</td></tr>
</table>

Note that no string in Σ^* has structure $S_{ab,\,ba}$ as its model.

Definition 3. $A = \langle D^A; \lhd, R_1^A, \ldots, R_n^A\rangle$ is a restriction of $B = \langle D^B; \lhd, R_1^B, \ldots, R_n^B\rangle$ iff $D^A \subseteq D^B$ and for each m-ary relation R_i, we have $R_i^A = \{(x_1 \ldots x_m) \in R_i^B \mid x_1, \ldots, x_m \in D^A\}$.

Informally, one identifies a subset A of the domain of B and strips B of all elements and relations which are not wholly within A. What is left is a restriction of B to A.

Definition 4. *Structure A is a subfactor of structure B $(A \sqsubseteq B)$ if A is connected, there exists a restriction of B denoted B', and there exists $h : A \to B'$ such that for all $a_1, \ldots a_m \in A$ and for all R_i in the model signature: if $h(a_1), \ldots h(a_m) \in B'$ and $R_i(a_1, \ldots a_m)$ holds in A then $R_i(h(a_1), \ldots h(a_m))$ holds in B'. If $A \sqsubseteq B$ we also say that B is a super-factor of A.*

In other words, properties that hold of the connected structure A also hold in a related way within B.

If $A \sqsubseteq B$ and $|A| = k$ then we say A is a k-subfactor of B. For all $w \in \Sigma^*$, and for any model M of Σ^*, let the subfactors of w be $\mathrm{Subfact}(M, w) = \{A \mid A \sqsubseteq M(w)\}$ and the k-subfactors of w be $\mathrm{Subfact}_k(M, w) = \{A \mid A \sqsubseteq M(w), |A| \leq k\}$. We also define $\mathrm{Subfact}(M, \Sigma^*)$ to be $\bigcup_{w \in \Sigma^*} \mathrm{Subfact}(M, w)$ and $\mathrm{Subfact}_k(M, \Sigma^*)$ to be $\bigcup_{w \in \Sigma^*} \mathrm{Subfact}_k(M, w)$. When M is understood from context, we write $\mathrm{Subfact}(w)$ instead of $\mathrm{Subfact}(M, w)$. We define the sets of superfactors $\mathrm{Supfact}(M, w)$ and $\mathrm{Supfact}(M, \Sigma^*)$ similarly.

Observe that $(\mathrm{Subfact}(M, w), \sqsubseteq)$ is a partially ordered set (poset). The next definition and lemma establishes that models of strings are principal elements of ideals and filters.

Definition 5 (Ideals). *A subset I of a poset is an Ideal if*

- *I is non-empty*

- *for every x in I, $y \leq x$ implies that y is in I*

- *for every x, y in I, there exists some element z in I, such that $x \leq z$ and $y \leq z$.*

The dual of an ideal is a filter.

Definition 6 (Filters). *A subset F of a poset is a filter iff*

- *F is non-empty*

- *for every x in F, $x \leq y$ implies that y is in F*

- *for every x, y in F, there exist some element z in F, such that $z \leq x$ and $z \leq y$.*

Definition 7 (Principal Ideals, Filters and Elements). *For any poset $\langle X, \leq\rangle$, the smallest filter containing $x \in X$ is a principal filter and x is the principal element of this filter. Similarly, the smallest ideal containing $x \in X$ is a principal ideal and x is the principal element of this ideal.*

Remark 1. *Given a model M of Σ^* and $k > 0$, $\mathrm{Subfact}_k(M, w)$ is a principal ideal in $\mathrm{Subfact}(M, \Sigma^*)$ whose principal element is $M(w)$. $\mathrm{Supfact}_k(M, w)$ is a principal filter in $\mathrm{Supfact}(M, \Sigma^*)$ whose principal element is $M(w)$. The empty structure $\langle \emptyset; \emptyset, \ldots \emptyset\rangle$ is a subfactor of every structure in $\mathrm{Subfact}(M, \Sigma^*)$.*

The next two propositions show how this representational perspective unifies the treatment of substrings and subsequences. They are subfactors under the successor and precedence models, respectively. A string $x = x_1 \cdots x_n$ is a substring of y iff there exists l, r such that $y = lxr$. String x is a subsequence of y iff there exists $v_0, v_1, \ldots v_n$ such that $w = v_0 x_1 v_1 \cdots x_n v_n$.

Proposition 1 (Substrings are subfactors under $M^{\lhd}$). *For all strings $x, y \in \Sigma^*$, x is a substring of y iff $M^{\lhd}(x) \sqsubseteq M^{\lhd}(y)$.*

Proof. Note that the result trivially holds for $x = \lambda$: we restrict ourselves to the case $x \neq \lambda$. Let $M^{\lhd}(x) = \langle D^x; \lhd, [R_\sigma^x]\rangle$ and $M^{\lhd}(y) = \langle D^y; \lhd, [R_\sigma^y]\rangle$

$(\Rightarrow)$. Suppose x is a substring of y: it exists l, r such that $y = lxr = \sigma_1 \ldots \sigma_{|l|} \sigma_{|l|+1} \cdots \sigma_{|l|+|x|} \sigma_{|l|+|x|+1} \cdots \sigma_{|l|+|x|+|r|}$. This implies that, for all i, $1 \leq i \leq |x|$, $d \in R^y_{\sigma_{|l|+i}}$ iff $d \in R^x_{\sigma_i}$. Thus, if we set the isomorphism ϕ to be such that $\phi(i) = |l| + i$ for $1 \leq i \leq |x|$, we have $\phi(M^{\lhd}(x))$ that is a restriction of $M^{\lhd}(y)$, and therefore $M^{\lhd}(x) \sqsubseteq M^{\lhd}(y)$ by definition.

($\Leftarrow$). Let y be the sequence of letters $\sigma_1 \ldots \sigma_{|y|}$ and suppose $M^\lhd(x) \sqsubseteq M^\lhd(y)$: there exists a isomorphism $\phi : \{1,\ldots,|x|\} \to \{1,\ldots,|y|\}$ such that $\phi(M^\lhd(x))$ is a restriction of $M^\lhd(y)$. This means that $\phi(D^x) \subseteq D^y$ and for all σ: $\phi(R^x_\sigma) = \{\phi(i) \in R^y_\sigma \mid \phi(i) \in \phi(D^x)\}$ (Definition 3). This implies that $x = \sigma_{\phi(1)} \ldots \sigma_{\phi(|x|)}$. Given that $\lhd = \{(i, i+1) \in D \times D\}$, we have $\phi(i+1) = \phi(i)+1$ and thus there exist l and r in Σ^* such that $y = l\sigma_{\phi(1)} \ldots \sigma_{\phi(|x|)} r = lxr$. $\qquad\square$

Proposition 2 (Subsequences are subfactors under $M^<$). *For all strings $x, y \in \Sigma^*$, x is a subsequence of y iff $M^<(x) \sqsubseteq M^<(y)$.*

Proof. We leave this proof to the Reader since it is of similar nature to the previous one. $\qquad\square$

4   Grammars, Languages, and Language Classes

Factors can define grammars, formal languages, and classes of formal languages. Usually a model signature provides the vocabulary for some logical language. Sentences in this logical language define sets of strings as follows. The language of a sentence ϕ is all and only those strings whose models satisfy ϕ. Within the regular languages, many well-known subregular classes can be characterized logically in this way (McNaughton and Papert, 1971; Rogers and Pullum, 2011; Rogers et al., 2013; Thomas, 1997).

Intuitively, the grammars we are interested in consist of a finite list of *forbidden* subfactors, whose largest size is bounded by k. Strings in the language of this grammar are those which do not contain any forbidden subfactors. In this way these grammars are like logical expressions which are "conjunctions of negative literals" (Rogers et al., 2013) where the negative literals are played by the the forbidden factors.

Each forbidden subfactor is a principal element of a filter and the language is all strings whose models are not in any of these filters. For each k, there is a class of languages including all and only those languages that can be defined in this way. For example, the Strictly k-Local (SL$_k$) and Strictly k-Piecewise languages can be defined in this way; they are languages which forbid finitely many substrings or subsequences, respectively (Garcia et al., 1990; Rogers et al., 2010). Formally:

Definition 8. *Let k be some positive integer, and M a model of Σ^* with signature Γ. A grammar G is a subset of $\mathtt{Subfact}_k(M, \Sigma^*)$. The language of G is $L(G) = \{w \in \Sigma^* \mid \mathtt{Subfact}_k(M, w) \cap G = \emptyset\}$. The class of languages $\mathscr{L}(M, k) = \{L \mid \exists G \subseteq \mathtt{Subfact}_k(M, \Sigma^*), L(G) = L\}$.*

The elements of G are principal elements of filters, and are called forbidden subfactors.

As an example, let $\Sigma = \{a, b, c\}$ and consider $G = \{M^\lhd(aa), M^\lhd(bb), M^\lhd(c)\}$. $L(G)$ includes the strings $(ab)^+$ and $(ba)^+$ and no other strings, because the substrings aa, bb, and c are all forbidden. This language belongs to $\mathscr{L}(M^\lhd, 2)$.

Proposition 3. *For each $w \in L(G)$ and each $g \in G$, $\mathtt{Subfact}(M, w)$ has a zero intersection with $\mathtt{Supfact}(g)$.*

Proof. Suppose there exists $A \in \mathtt{Subfact}_k(\Sigma^*)$ such that $A \sqsubseteq M(w)$ and $g \sqsubseteq A$. This implies that $g \sqsubseteq M(w)$ and thus that $\mathtt{Subfact}_k(M, w) \cap G \neq \emptyset$ which contradicts Definition 8. $\qquad\square$

In other words, the principal ideal of $M(w)$ is disjoint from the principal filters of the elements of G.

5   Grammatical Entailments

Given a grammar G, we call a subfactor s in $\mathtt{Subfact}(\Sigma^*)$ *ungrammatical* if it belongs to a principal filter of any element of G. Subfactors that are not ungrammatical are called *grammatical*. Lemma 14 ensures that grammaticality is downward entailing, in the sense that if a model of the word $M(w)$ is not contained in the principal filters of the elements of the grammar, then neither are the subfactors of $M(w)$. But it also ensures that ungrammaticality is upward entailing: if a model of the word $M(w)$ belongs to the principal filters of the elements of the grammar, then all of the superfactors of $M(w)$ in that filter are likewise contained.

In this way, the ideals and filters within a a particular model noted above give rise to these entailment properties of grammaticality with respect to the hypothesis space. If the learner constructs filters, then the grammar G will allow structures such that language membership is downward entailing with respect to the grammar G, and language non-membership is upward entailing with respect to the grammar G.

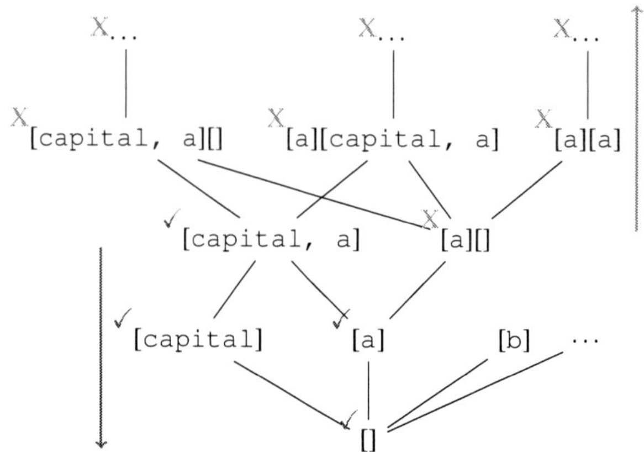

Figure 3: The Structure ideals(blue) and filters(red) for a capitalized letter model.

5.1 Example: Text Capitalization

As an example, consider capitalized letters as discussed above. In an unconventional word model, each capital letter at some position x is represented as satisfying one of the relations $R \in \{a(x), b(x), \ldots, z(x)\}$ as well as the unary relation $\texttt{capital}(x)$. Thus the relation $a(x)$ is true of both lowercase a and uppercase A, but $a(x) \wedge \texttt{capital}(x)$ is only true of uppercase A. Note also that in this model no position x of a structure can satisfy both predicates $a(x)$ and $b(x)$. We return to this point in §7.

Figure 2 showcases the relationship among these structures under a model M. The structure for A, [capital, a], contains as subfactors [capital], [a], [], and the empty structure (not shown). The empty structure is a subfactor of [], and [] in turn is a subfactor of [capital] and [a]. The subfactor [a] contains the subfactor [], the domain element with no relations, but has superfactors [capital, a], which has one domain element and two relations, and [a][], which has two domain elements, and the first satisfying the property a. Subfactors and superfactors are listed above and below each other, respectively, with lines between them. Members of one ideal are noted with a blue checkmark, and members of a filter are noted by a red asterisk.

Applying this to the example in Figure 3, if the structure [capital, a] is grammatical, then all of its subfactors, such as [capital] and [a], and [] are grammatical. Since those are grammatical, each of their subfactors is also grammatical, which in this case is just [0], shown in blue in Figure 3. Conversely, if the structure [a][] is known to be ungrammatical, then any structure which has it as a subfactor is also ungrammatical (in this example, [capital, a][], shown in Red in Figure 3. To see the importance, consider a string with only lowercase letters. In a connected model, the grammar would ban 26 forbidden factors (A,B,C,...), but the "capital" model bans just one, [capital].

5.2 Example: Long Distance Linguistic Dependencies

As another example, sequences of speech sounds as mentioned earlier may be decomposed into binary features based on their phonetic properties like anteriority ($\pm\texttt{ant}$ — whether it occurs in the anterior of the vocal tract), stridency ($\pm\texttt{str}$ — whether it produces a high-intensity fricative noise), or voicing ($\pm\texttt{voi}$ — whether it activates the vocal chords), among others (Hayes, 2009). Each sound at some position x is represented as satisfying relations $R \in \{\pm\texttt{voi}(x), \pm\texttt{str}(x), \ldots, \pm\texttt{ant}(x)\}$. Thus the relation $+\texttt{str}(x)$ is true of both the sound s as in the first sound of "sue" and ʃ, as in "shoe", but $+\texttt{str}(x) \wedge -\texttt{ant}(x)$ is only true of ʃ.

Note also that in this model no position x of a structure can satisfy both predicates $+\texttt{str}(x)$ and $-\texttt{str}(x)$. We return to this point in §7 below. We again use square brackets to delimit the domain elements and write the unary features within them, so a model representation like $\begin{bmatrix}+\texttt{str}\\+\texttt{ant}\end{bmatrix}\begin{bmatrix}+\texttt{str}\\-\texttt{ant}\end{bmatrix}$ has the following visual representation:

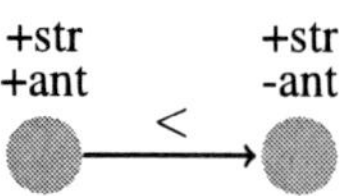

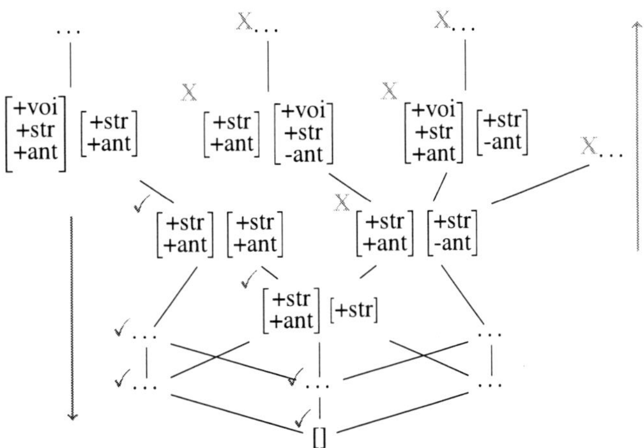

Figure 4: Structure ideals(blue) and filters(red) for a phonological precedence model.

To ease the exposition, we will use square brackets to delimit the domain elements and write the unary relations within them instead of specifying the model in mathematical detail. In an unconventional subsequence word model, then, one possible structure of the subsequence s...ʃ is written $\begin{bmatrix} +\texttt{str} \\ +\texttt{ant} \end{bmatrix} \begin{bmatrix} +\texttt{str} \\ -\texttt{ant} \end{bmatrix}$.

In many languages, the presence of certain segments is dependent on the presence of another segment. In Samala, subsequences like s...s are allowed but s...ʃ are not, so words like ha**s**xintilawa**s** are allowed but words like ha**s**xintilawaʃ are not (Hansson, 2010). In an unconventional model, banning structures of the form [+str][+str] is insufficient, since all these segments share that stridency property, while a structure like $\begin{bmatrix} +\texttt{str} \\ +\texttt{ant} \end{bmatrix} \begin{bmatrix} +\texttt{str} \\ -\texttt{ant} \end{bmatrix}$ will distinguish them, since they disallow stridents which disagree on the $\pm\texttt{ant}(x)$ relations. The structure [+ant][-ant] however, is insufficient, since consonants like p,b,m have that feature, and would incorrectly ban acceptable strings. To see the importance, a conventional string model must ban multiple sibilant factors sʃ,zʃ,sʒ,zʒ, while an unconventional model must just ban one, $\begin{bmatrix} +\texttt{str} \\ +\texttt{ant} \end{bmatrix} \begin{bmatrix} +\texttt{str} \\ -\texttt{ant} \end{bmatrix}$

Figure 4 showcases the relationship among these structures under a precedence model $M^<$. The structure for $\begin{bmatrix} +\texttt{str} \\ +\texttt{ant} \end{bmatrix}$[+str] contains as subfactors (among others) [+str][+str], [+str][], [], and the empty structure (not shown). The empty structure is a subfactor of [], and [] in turn is a subfactor of [+ant] and [-str], and so on.

If the structure $\begin{bmatrix} +\texttt{str} \\ +\texttt{ant} \end{bmatrix} \begin{bmatrix} +\texttt{str} \\ +\texttt{ant} \end{bmatrix}$ is grammatical, then all of its subfactors, are grammatical, and so are their subfactors, in turn. Conversely, if the structure $\begin{bmatrix} +\texttt{str} \\ +\texttt{ant} \end{bmatrix} \begin{bmatrix} +\texttt{str} \\ -\texttt{ant} \end{bmatrix}$ is known to be ungrammatical, then any structure which has it as a subfactor is also ungrammatical (for example, $\begin{bmatrix} +\texttt{voi} \\ +\texttt{str} \\ +\texttt{ant} \end{bmatrix} \begin{bmatrix} +\texttt{str} \\ -\texttt{ant} \end{bmatrix}$, where the first segment is also voiced +voi), shown in Red in Figure 4.

The structure filters give the learner an advantage when confronting hypothesis spaces under a particular model. In particular, it allows the learner to prune vast swathes of the hypothesis space as it reaches for principal elements of features. If a learner identifies one structure as being grammatical, the learner may infer that all of its subfactors are also grammatical and not have to consider them. Alternatively, if the learner knows a structure is ungrammatical, it may infer that the ideals above it are also ungrammatical.

Generally, these reductions can be exponential: an alphabet of size 2^n can be represented with n unary relations in the model signature. However, this exponential reduction does not necessarily make learning any easier. The reason for this is that the size of $\texttt{Subfact}_k(M, \Sigma^*)$ equals $\sum_{i=1}^{k}(2^n)^i$ where n is the number of unary relations. Since a grammar is defined as a subset of $\texttt{Subfact}_k(M, \Sigma^*)$, the number of considered grammars is thus very large. Therefore, the problem of how to search this space effectively is paramount.

6 The Learning Problem

For some M, k, is $\mathscr{L}(M,k)$ learnable from positive data? The short answer is Yes (Heinz, 2010; Heinz et al., 2012). The solution presented in these papers can be thought of as using the function $\texttt{Subfact}_k(M,w)$ to identify permissible k-factors in words w in the positive data. The k-factors that are not permissible are forbidden. With sufficient positive data, such a learning algorithm will converge to a grammar that generates any target language in the class. While this solution is sound in theory, when the space of k-factors is very large, it is not practical. Here, we make clear the problem the learning algorithm solves.

We state the learning problem not in terms of converging to a correct grammar in the limit as previously studied, but instead of returning an 'adequate' grammar given a finite positive sample. Determining what counts as an adequate grammar is what (De Raedt, 2008) calls a Quality Criterion.

Definition 9 (The Learning Problem). *Fix* Σ, *model M, and positive integer k. For any language* $L \in \mathscr{L}(M,k)$ *and for any finite* $D \subseteq L$, *return a grammar G such that*

1. *G is consistent, that is, it covers the data:* $D \subseteq L(G)$;

2. $L(G)$ *is a smallest language in* $\mathscr{L}$ *which covers the data: so for all* $L \in \mathscr{L}$ *where* $D \subseteq L$, *we have* $L(G) \subseteq L$; *and*

3. *G includes structures S that are restrictions of structures S' included in other grammars G' that also satisfy (1) and (2): for all G' satisfying the first two criteria for all $S' \in G'$, there exists $S \in G$ such that $S \sqsubseteq S'$.*

The first criterion is self-explanatory. The second criterion is motivated by Angluin's (1980) analysis of identification in the limit. The third criterion requires that the grammar contain the most "general" subfactors. An example will help illustrate this criterion.

Consider again the grammar $G = \{M^{\triangleleft}(aa), M^{\triangleleft}(bb), M^{\triangleleft}(c)\}$ with $\Sigma = \{a,b,c\}$. $L(G)$ is the same as $L(H)$ where $H = \{M^{\triangleleft}(aa), M^{\triangleleft}(bb), M^{\triangleleft}(ac), M^{\triangleleft}(bc), M^{\triangleleft}(cc), M^{\triangleleft}(ca), M^{\triangleleft}(cb)\}$. In H all the forbidden subfactors are of size 2, whereas G encapsulates all of the 2-factors in H which include c with a single 1-factor $M^{\triangleleft}(c)$. Both grammars G and H may satisfy criteria (1) and (2) but H would not satisfy criterion (3) because of G.

7 A Bottom-Up Learning Algorithm

(De Raedt, 2008) identifies two directions of inference: specific-to-general (i.e., 'top-down') and general-to-specific (i.e., 'bottom-up'). Since we are trying to find the most general subfactors, top-down inference has the potential to consider exponentially many more subfactors than bottom-up inference. It makes mores sense to traverse bottom-up, that is, from the most general subfactors possible to the most specific. Additionally, once a subfactor is identified as an element of the grammar, none of its superfactors (elements of its principal filter) need to be considered further.

A bottom-up learner is shown in Algorithm 1. Its input is a positive data sample D and an integer k that identifies the upper bound on the size of the subfactors.

Data: positive sample D, empty structure s_0, max constraint size k
Result: G, a set of constraints
$Q \leftarrow \{s_0\}$;
$G \leftarrow \emptyset$;
$V \leftarrow \emptyset$;
while $Q \neq \emptyset$ **do**
    $s \leftarrow Q.\texttt{dequeue}()$;
    $V \leftarrow V \cup \{s\}$;
    if $\exists x \in D$ *such that* $s \sqsubseteq x$ **then**
        $S \leftarrow \texttt{NextSupFact}(s)$;
        $S' \leftarrow \{s \in S \mid |s| \leq k \wedge (\neg \exists g \in G)[g \sqsubseteq s] \wedge s \notin V\}$;
        $Q.\texttt{enqueue}(S')$;
    end
    else
        $G \leftarrow G \cup \{s\}$;
    end
end
return G;

Algorithm 1: Bottom-up learning algorithm for lattice-structured constraints

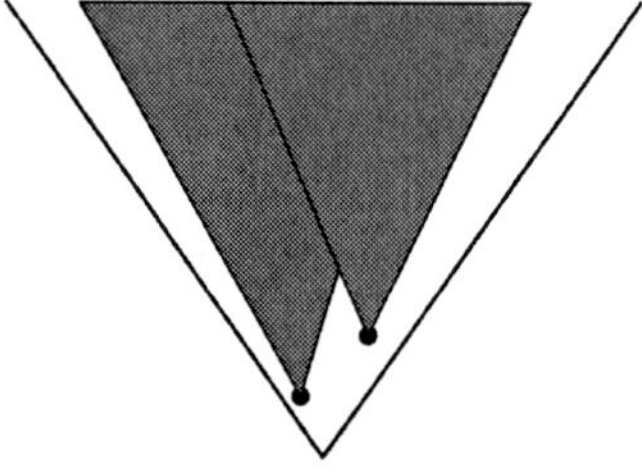

Figure 5: Pruning the hypothesis space

The algorithm makes use of a queue Q, which is initialized to contain just the empty structure s_0. It also initializes two empty sets: G, the grammar that will ultimately be returned, and V, the set of 'visited subfactors'. The subfactors in Q are considered one at a time, in order, and as each subfactor s is considered it is added to V. If s is not a subfactor of the model of any word in the positive sample D (i.e., not contained by any data point in D), then it is added to the grammar G.

If s is a subfactor of the sample, it is sent to the function `NextSupFact`, which returns a set of *least* superfactors for s. For concreteness, `NextSupFact(s)` may be defined formally as follows:

$$\texttt{NextSupFact}(s) = \{S \in \texttt{Subfact}_k(\Sigma^*) \mid s \sqsubseteq S, \neg \exists S'[s \sqsubseteq S' \sqsubseteq S]\}.$$

Practically `NextSupFact` will be defined constructively so that each subfactor in $\texttt{Subfact}_k(\Sigma^*)$ is constructed only once as needed. Thus, not only will it not be needed to store the whole set $\texttt{Subfact}_k(\Sigma^*)$ in memory, but the set V may be excluded from the algorithm as well.

This set of superfactors is then filtered by the following criteria: they must be smaller than $k+1$, they must contain no element of G as a subfactor, and they must not have been previously considered (i.e., they cannot be in V). Those structures that survive this filter are added to Q. This procedure continues until there are no more structures left to consider in Q.

Theorem 1. *For any $L \in \mathscr{L}(M,k)$, and any finite set $P \subseteq L$ provided as input to Algorithm 1, it returns a grammar G satisfying Definition 9.*

Proof. Consider any $x \in D$. Algorithm 1 only adds elements to G that are not subfactors of x, so $x \notin \texttt{Supfact}(G)$. Thus $x \in L(G)$ and $D \subseteq L(G)$, satisfying Condition (1).

Consider any $L' \in \mathscr{L}$ with $D \subseteq L'$. To show $L = L(G) \subseteq L'$, consider any $w \in L$. Then $\texttt{Subfact}(w) \subseteq \texttt{Subfact}(D)$ and $\texttt{Subfact}(D) \subseteq \texttt{Subfact}(L')$ since $D \subseteq L$. Then $\texttt{Subfact}(w) \subseteq \texttt{Subfact}(L')$. Hence, $w \in L'$, and so $L \subseteq L'$, satisfying Condition (2).

For condition (3), we use the fact that elements in the grammar G were in Q at some point. Suppose s, s' are subfactors such that $s \in G$, $s' \sqsubseteq s$, and $(\neg \exists x \in D)[s' \sqsubseteq M(x)]$. Since $s \in G$, then at some point $s \in Q$.

If $s' \sqsubseteq s$ then s' will be added to Q before s is generated by `NextSupFact`. Because Q is a queue, s' will also be removed from Q before s is generated by `NextSupFact`. Since s' is not contained by any $M(x)$ with $x \in D$, it will be added to G. When s is generated by `NextSupFact`, it will

not pass the filter because it fails the second criterion since $s' \sqsubseteq s$ and $s' \in G$. Then s is never added to Q, and therefore $s \notin G$, contra our original assumption. Thus Condition (3) is satisfied. $\qquad\square$

One aspect of the algorithm to highlight is that when a subfactor g is added to G, it is not added to Q. Consequently, `NextSupFact(g)` is never added to Q. In this way, finding elements of G prunes the remainder of the space to be searched (see figure 5). In general, it is not the case that every element in the principal filter of g will not be generated by `NextSupFact` since some of these elements may belong to `NextSupFact(x)` for other subfactors x on the Q. We expect subfactors on the 'border' of $\texttt{Supfact}(g)$ to be generated in this way (and then they are filtered out). This pruning, especially when the subfactors are quite general, can significantly reduce the remaining space to be traversed.

In regard to efficiency, in the worst case, the elements of G are all very specific subfactors and are greatest elements of $\texttt{Subfact}_k(\Sigma^*)$. In this case, every subfactor $\texttt{Subfact}_k(\Sigma^*)$ will be added to Q and the time complexity is thus exponential. However, we are primarily interested in the case when $\texttt{Subfact}_k(D)$ are a small proportion of $\texttt{Subfact}_k(\Sigma^*)$. This constitutes an example of data sparsity. In this case, we believe the elements of the target grammar will be much 'lower' in the partial order and thus will be found much more quickly. Determining what conditions on $\texttt{Subfact}_k(D)$ and $\texttt{Subfact}_k(\Sigma^*)$ result in a polynomial time run in the size of D is a focus of current research activity.

Another area of active research is developing a recipe for the `NextSupFact` function for models with a successor or precedence order relation and arbitary unary relations. The basic idea underlying the bottom-up algorithm is to develop a spanning tree for the poset $\texttt{Subfact}(\Sigma^*)$ and to traverse this tree in a breadth-first manner. The function `NextSupFact` helps control this search. Ideally, `NextSupFact` would only generate each subfactor once, which obviates the need to store visited subfactors in V. This can be accomplished to some extent in different ways. For incompatible unary relations, like a and b in our capitalization example, `NextSupFact` can be defined to prevent adding property a to a position that already satisfies property b.

For compatible unary relations, like a and

capital in our capitalization example, an ordering over the unary relations such as a < b < capital can help eliminate generating the same subfactor in different ways. For example, if NextSupFact is defined to only add 'lesser' unary relations to positions that already have them then it would only output [capital,a] given the subfactor [a] as input. On the other hand, when given as input the subfactor [capital], it could not add any unary relation to this position.

8 Conclusion

In this paper, we considered the problem of learning formal languages defined as the complement of the union of finitely many principal filters, whose principal elements make up the grammar. This is one way to characterize the Strictly k-Local and Strictly k-Piecewise languages, but the generalization here lets us consider enriched representations of strings where different elements in a string can be said to share properties. it also lets us learn the shortest forbidden substrings in SL_k (Ron et al., 1996) This is useful in many applications where domain-specific knowledge is available and should be taken advantage of. Such enriched representations, however, have a drawback. The number of subfactors is large which makes identifying the principal elements of the filters difficult. This paper showed that the partial ordering of the subfactors motivates a bottom-up learning algorithm which finds the least subfactors whose filters do not include the positive data.

Acknowledgments

We would like to thank James Rogers for very helpful discussion on the notion of subfactor. This work was supported by NIH grant #R01HD87133-01 to JH.

References

Dana Angluin. 1980. Inductive inference of formal languages from positive data. *Information Control*, 45:117–135.

J. Richard Büchi. 1960. Weak second-order arithmetic and finite automata. *Mathematical Logic Quarterly*, 6(1-6):66–92.

Luc De Raedt. 2008. *Logical and Relational Learning*. Springer-Verlag Berlin Heidelberg.

Herbert B. Enderton. 2001. *A Mathematical Introduction to Logic*, 2nd edition. Academic Press.

Pedro Garcia, Enrique Vidal, and José Oncina. 1990. Learning locally testable languages in the strict sense. In *Proceedings of the Workshop on Algorithmic Learning Theory*, pages 325–338.

Saul Gorn. 1967. Explicit definitions and linguistic dominoes. In *Systems and Computer Science*, pages 77–115, Toronto. University of Toronto Press.

Gunnar Hansson. 2010. *Consonant Harmony: Long-Distance Interaction in Phonology*. Number 145 in University of California Publications in Linguistics. University of California Press, Berkeley, CA. Available on-line (free) at eScholarship.org.

Bruce Hayes. 2009. *Introductory Phonology*. Wiley-Blackwell.

Jeffrey Heinz. 2010. String extension learning. In *Proceedings of the 48th Annual Meeting of the Association for Computational Linguistics*, pages 897–906, Uppsala, Sweden. Association for Computational Linguistics.

Jeffrey Heinz, Anna Kasprzik, and Timo Kötzing. 2012. Learning with lattice-structured hypothesis spaces. *Theoretical Computer Science*, 457:111–127.

Colin de la Higuera. 2010. *Grammatical Inference: Learning Automata and Grammars*. Cambridge University Press.

Leonid Libkin. 2004. *Elements of Finite Model Theory*. Springer.

Robert McNaughton and Seymour Papert. 1971. *Counter-Free Automata*. MIT Press.

James Rogers. 2003. Syntactic structures as multi-dimensional trees. *Research on Language and Computation*, 1(3-4):265–305.

James Rogers, Jeffrey Heinz, Gil Bailey, Matt Edlefsen, Molly Visscher, David Wellcome, and Sean Wibel. 2010. On languages piecewise testable in the strict sense. In *The Mathematics of Language*, volume 6149 of *Lecture Notes in Artifical Intelligence*, pages 255–265. Springer.

James Rogers, Jeffrey Heinz, Margaret Fero, Jeremy Hurst, Dakotah Lambert, and Sean Wibel. 2013. Cognitive and sub-regular complexity. In *Formal Grammar*, volume 8036 of *Lecture Notes in Computer Science*, pages 90–108. Springer.

James Rogers and Geoffrey K. Pullum. 2011. Aural pattern recognition experiments and the subregular hierarchy. *Journal of Logic, Language and Information*, 20:329–342.

Dana Ron, Yoram Singer, and Naftali Tishby. 1996. The power of amnesia: Learning probabilistic automata with variable memory length. *Machine Learning*, 25(2-3):117–149.

Kristina Strother-Garcia, Jeffrey Heinz, and Hyun Jin Hwangbo. 2016. Using model theory for grammatical inference: a case study from phonology. In *Proceedings of The 13th International Conference on Grammatical Inference*, volume 57 of *JMLR: Workshop and Conference Proceedings*, pages 66–78.

Wolfgang Thomas. 1997. Languages, automata, and logic. In *Handbook of Formal Languages*, volume 3, chapter 7. Springer.

Mai Ha Vu, Ashkan Zehfroosh, Kristina Strother-Garcia1, Michael Sebok, Jeffrey Heinz, and Herbert G. Tanner. 2018. Statistical relational learning with unconventional string models. *Frontiers in Robotics and AI*.

Maximum Likelihood Estimation of Factored Regular Deterministic Stochastic Languages

Chihiro Shibata
School of
Computer Science
Tokyo University of Technology
shibatachh@stf.teu.ac.jp

Jeffrey Heinz
Department of Linguistics &
Institute for Advanced Computational Science
Stony Brook University
jeffrey.heinz@stonybrook.edu

Abstract

This paper proves that for every class C of stochastic languages defined with the co-emission product of finitely many probabilistic, deterministic finite-state acceptors (PDFA) and for every data sequence D of finitely many strings drawn i.i.d. from some stochastic language, the Maximum Likelihood Estimate of D with respect to C can be found efficiently by locally optimizing the parameter values. We show that a consequence of the co-emission product is that each PDFA behaves like an independent factor in a joint distribution. Thus, the likelihood function decomposes in a natural way. We also show that the negative log likelihood function is convex. These results are motivated by the study of Strictly k-Piecewise (SP_k) Stochastic Languages, which form a class of stochastic languages which is both linguistically motivated and naturally understood in terms of the co-emission product of certain PDFAs.

1   Introduction

Stochastic languages are probability distributions over all possible strings of finite length. A class C of stochastic languages is often defined parametrically: an assignment of values to the parameters uniquely determines some stochastic language L in C and thus the probabilities that L assigns to strings. An important learning criterion for a class of stochastic languages C is whether there is an algorithm which reliably returns a Maximum-Likelihood Estimate (MLE) of an observed data sample D. The MLE is the parameter values which maximize the probability of D with respect to C.

This paper focuses on *regular deterministic* stochastic languages. These are stochastic languages that can be defined with a probabilistic, deterministic, finite-state acceptors (PDFA).

The problem of finding the MLE, however, is not only about some single stochastic language L, but also about the *class* of stochastic languages that L belong to. It is well-understood that each PDFA $\mathcal{M}$ naturally defines a class of stochastic languages $C_{\mathcal{M}}$ because the transitional probabilities in the PDFA provide a range of possible parameter values, as we explain in detail in section 2. In this case, it is well-understood how to find the MLE of a sequence of strings drawn i.i.d. from L with respect to $C_{\mathcal{M}}$ (Vidal et al., 2005a,b). This paper is concerned with finding the MLE for *different* classes of stochastic languages.

In particular, we consider the case where C is defined by the range of parametric values over *finitely many* PDFA $\mathcal{A} = \{\mathcal{M}_1 \ldots \mathcal{M}_K\}$, whose *co-emission product* determines the probabilities each $L \in C$ assigns to strings. Essentially, the co-emission product of these PDFAs *factor* the probabilities each $L \in C$ assigns to strings. Each L is a complex joint distribution, and each PDFA $\mathcal{M}_j$ represents a 'more basic' regular stochastic language whose parameter values independently contribute to L. At a high level, the problem we are considering is like those addressed with Bayesian networks and Markov random fields, where complex probability distributions decompose into simpler factors (Bishop, 2006; Koller and Friedman, 2009). We refer to the classes C we study in this paper as factored, regular, probabilistic, and deterministic (FRPD).

The main result is to show how the parameters of a FRPD class C can be efficiently updated to find those parameter values which maximize the likelihood of the observed sequences (Theorem 2). We also show directly that each negative log likelihood associated with each FRPD class C is convex (Theorem 3). Together these results imply that the efficient method we present for updating the parameter values will yield the MLE.

There are several reasons for being interested in such factored classes C. Perhaps the most important from our perspective is expressed by Koller and Friedman (p. 1134) "The ability to exploit structure in the distribution is the basis for providing a compact representation of high-dimensional ... probability spaces." In our case, the size of the representation of the class given by $\mathcal{A} = \{\mathcal{M}_1 \ldots \mathcal{M}_K\}$ is simply the sum of the size of each M_j. In contrast, the representation of the class given by the co-emission product is in the worst case the product of the sizes of each M_j. One direct benefit of this is that the number of parameters is reduced, which makes it possible to more accurately estimate them with less data. Other advantages discussed by Koller and Friedman, such as modularity, we return to in the discussion in the conclusion.

There are also linguistic reasons to be interested in FRPD classes. The Strictly Piecewise (SP) class of languages encode certain types of long-distance dependencies found in natural languages. For example, SP languages can express generalizations like "at most one b per string" and "no b may follow an a" (Rogers et al., 2010). Generalizations with this formal character are known to occur in the phonologies of the world's languages (Heinz, 2010a; Rogers et al., 2013; Heinz, 2014, 2018). As Rogers et al. (2010) explain, Strictly Piecewise languages are characterized by the intersection product of finitely many deterministic finite-state acceptors (DFA). Heinz and Rogers (2010) used this characterization and the co-emission product to define the class of Strictly Piecewise stochastic languages because they were interested in the learnability of long-distance dependencies in natural languages probabilistically. They presented a learning algorithm for a class of SP stochastic languages and claimed (p. 894) that it returns the MLE.

This results in this paper can be seen as providing a more generalized, more meaningful, and more rigorous proof of their basic claim. Theorem 2 establishes how to update the parametric values which locally optimize the model of *any* FRPD class. Theorem 3 shows the negative log likelihood function of *any* FRPD class is convex, so there is in fact only one set of optimal parametric values for any sequence of data. Furthermore, we prove these results in terms of the standard definition of co-emission product, and not the variant used in Heinz and Rogers (2010). (While the results here work for both, we only prove the standard case.) These general results make it possible to explore not only the learning of SP_k stochastic languages, but also *any* finite combination of PDFAs that characterize different kinds of local and non-local dependencies which can be expressed with regular grammars. We return to this issue in the discussion.

To our knowledge, such results for FRPD classes have not been previously discussed in the literature. One reason for this is that much work on natural language processing uses probabilistic *non-deterministic* automata. These describe the same class of stochastic languages as Hidden Markov Models (HMMs) (Vidal et al., 2005a,b). Non-determinism can make a big difference when it comes to parsing and learning. In a deterministic model $\mathcal{M}$, each string w can be associated with at most one path through $\mathcal{M}$, whereas in non-deterministic $\mathcal{M}$, there can be infinitely many paths for w. This is one reason why methods used for learning HMM are not guaranteed to return a MLE. Since the states are 'hidden' one uses methods like Expectation Maximization, which may converge to a local optimum that is not a global optimum (Jurafsky and Martin, 2008; Heinz et al., 2015).

On the other hand, we are showing that, by carefully choosing the class of stochastic languages C—which the MLE which is to be found will be 'with respect to'—we can exploit the structure we assume to be present to guarantee we find a MLE. This paper takes one step in establishing the theoretical soundness of this approach.

Finally, one reviewer commented that these results may follow from fundamental theorems in the literature on probabilistic graphical models (Koller and Friedman, 2009). Regardless of whether this is true, the correctness of the proofs here stand. Also, the general results of Bayesian networks and Markov random fields say nothing about the concrete forms of the algorithm for obtaining the MLE with respect to a FRPD class C given data D, and similarly for its time complexity. Malouf (2002) makes a similar point, writing "While all parameter estimation algorithms we will consider take the same general form, the method for computing the updates ... differs substantially." Nonetheless, how probabilistic graphical models relate to this line of research ought to be made clear.

The remainder of the paper is organized as follows. In section 2 we review languages, stochastic languages, deterministic finite-state acceptors and probabilistic versions thereof, the intersection and co-emission products, and the statement of the learning problem. Before presenting our main results, section 3 defines Strictly Piecewise (stochastic) languages, which provide a running example to illustrate the main results, which are presented in section 4. The computational complexity of the updates are analyzed in section 5 and section 6 concludes.

2 Preliminaries

2.1 Sets of Strings

Σ denotes a finite set of symbols and Σ^k, $\Sigma^{\leq k}$, and Σ^* denote all strings over this alphabet of length k, of length less than or equal to k, and of any finite length, respectively. λ denotes the empty string. The length of a string w is written $|w|$. The prefixes of a string w are $\mathrm{Pref}(w) = \{v \mid \exists u \in \Sigma^*, vu = w\}$. A string $w = \sigma_1 \ldots \sigma_n$ is a *subsequence* of a string v if and only if $v \in \Sigma^* \sigma_1 \Sigma^* \ldots \Sigma^* \sigma_n \Sigma^*$, in which case we write $w \sqsubseteq v$.

A *language* L is a subset of Σ^*. The *complement* of a language L, denoted $\overline{L}$ is Σ^*/L. The *shuffle ideal* of w is the language of all strings containing w as a subsequence:

$$SI(w) = \{v \mid w \sqsubseteq v\}.$$

A *stochastic language* L is a probability distribution over Σ^*. The probability P of word w with respect to L is written $P_L(w) = p$. Thus, all stochastic languages L satisfy

$$\sum_{w \in \Sigma^*} P_L(w) = 1.$$

2.2 Probabilistic Deterministic Finite-state Acceptors

A *Deterministic Finite-state Acceptor* (DFA) is a tuple $\mathcal{M} = \langle Q, \Sigma, q_0, \delta, F \rangle$ where Q is the state set, Σ is the alphabet, q_0 is the start state, δ is a deterministic transition function with domain $Q \times \Sigma$ and codomain Q, and F is the set of accepting states. Let $\delta^* : Q \times \Sigma^* \to Q$ be the (partial) path function of $\mathcal{M}$. When discussing partial functions, the notation $\uparrow$ and $\downarrow$ indicates that the function is not defined, respectively, is defined, for particular arguments. Thus

$\delta^*(q, w)$ is the (unique) state reachable from state q via the sequence w, if any, or $\delta^*(q, w)\uparrow$ otherwise. The language recognized by a DFA $\mathcal{M}$ is $L(\mathcal{M}) \overset{\text{def}}{=} \{w \in \Sigma^* \mid \delta^*(q_0, w)\downarrow \in F\}$.

A *Probabilistic Deterministic Finite-state Acceptor* (PDFA) is a tuple $\mathcal{M} = \langle Q, \Sigma, q_0, \delta, F, T \rangle$ where Q, Σ, q_0, and δ are the same as with DFA, and F and T are partial functions representing the final-state and transition probabilities. In particular, $T : Q \times \Sigma \to \mathbb{R}^+$ and $F : Q \to \mathbb{R}^+$ such that

$$\text{for all } q \in Q, \ F(q) + \sum_{\sigma \in \Sigma} T(q, \sigma) = 1. \quad (1)$$

A PDFA $\mathcal{M}$ generates a stochastic language $L(\mathcal{M})$. If it exists, the unique *path* for a word $w = \sigma_0 \ldots \sigma_N$ belonging to Σ^* through a PDFA is a sequence $\langle (q_0, \sigma_0), (q_1, \sigma_1), \ldots, (q_N, \sigma_N) \rangle$, where $q_{i+1} = \delta(q_i, \sigma_i)$. The probability a PDFA assigns to w is obtained by multiplying the transition probabilities along w's path if it exists with the final probability, and zero otherwise. So $P_{L(\mathcal{M})}(w) =$

$$\left(\prod_{i=0}^{N} T(q_i, \sigma_i) \right) \cdot F\big(\delta(q_N, \sigma_N)\big)$$

$$\text{if } \delta^*(q_0, w)\downarrow \text{ and 0 otherwise} \quad (2)$$

A probability distribution is *regular deterministic* iff there is a PDFA which generates it. We sometimes write $\mathcal{M}(w)$ instead of $P_{L(\mathcal{M})}(w)$.

The *structural components* of a PDFA $\mathcal{M}$ are its states Q, its alphabet Σ, its transitions δ, and its initial state q_0. By *structure* of a PDFA, we mean its structural components. The structure of each PDFA $\mathcal{M}$ defines a class of stochastic languages given by the possible instantiations of T and F satisfying Equation 1. These distributions have at most $|Q| \cdot (|\Sigma| + 1)$ independent parameters (since for each state there are $|\Sigma|$ possible transitions plus the possibility of finality.)

2.3 The co-emission product

The *intersection product* of K DFAs $\mathcal{M}_1 \ldots \mathcal{M}_K$ is given by the standard construction over the state space $Q_1 \times \ldots \times Q_K$ (Hopcroft et al., 2001). We write $\bigotimes_{1 \leq j \leq K} \mathcal{M}_j = \mathcal{M} = \langle Q, \Sigma, q_0, \delta, F \rangle$ where $Q = Q_1 \times \ldots \times Q_K$, $q_0 = \langle q_{01}, \ldots q_{0K} \rangle$. For all $\langle q_1, \ldots q_K \rangle \in Q$ and $\sigma \in \Sigma$, $\delta(\langle q_1, \ldots q_K \rangle, \sigma) = \langle q_1', \ldots q_K' \rangle$ if and only if $\delta_1(q_1, \sigma) = q_1', \ldots \delta_K(q_K, \sigma) = q_K'$. Finally, let $F = F_1 \times \ldots \times F_K$ It is well-known that $L(\bigotimes_{1 \leq j \leq K} \mathcal{M}_j) = \bigcap_{1 \leq j \leq K} L(\mathcal{M}_j)$.

The *co-emission product* of K PDFAs $\mathcal{M}_1 \dots \mathcal{M}_K$ is also given by a construction over the state space $Q_1 \times \dots \times Q_K$. The probability that σ is co-emitted from $\langle q_1, \dots, q_K \rangle$ in $Q_1 \times \dots] \times Q_K$ is the product of the probabilities of its emission at each $q_j \in Q_j$. Let $\mathsf{CoT}(\langle \sigma, q_1 \dots q_K \rangle) = \prod_{j=1}^{K} T_j(q_j, \sigma)$. Similarly, the probability that a word simultaneously ends at $q_1 \in Q_1, \dots q_K \in Q_K$ is

$$\mathsf{CoF}(\langle q_1 \dots q_K \rangle) = \prod_{j=1}^{K} F_j(q_j).$$

Finally, for $q = \langle q_1 \dots q_K \rangle$, let

$$Z(q) = \mathsf{CoF}(q) + \sum_{\sigma \in \Sigma} \mathsf{CoT}(\langle \sigma, q \rangle)$$

be the *normalization term*. Next we define the co-emission product.

Definition 1 (Co-emission Product) *For* $\mathcal{A} = \{\mathcal{M}_1, \dots \mathcal{M}_K\}$, *let* $\bigotimes \mathcal{A} = \langle Q, \Sigma, q_0, \delta, F, T \rangle$ *where*

1. *$Q, q_0,$ and δ are defined as with DFA product; and*

2. *For all $q \in Q$ and $\sigma \in \Sigma$:*
$$F(q) = \frac{\mathsf{CoF}(q)}{Z(q)}$$
and
$$T(q, \sigma) = \frac{\mathsf{CoT}(\sigma, q)}{Z(q)}.$$

In other words, the numerators of T and F are defined to be the co-emission probabilities and division by Z ensures that co-emission product $\bigotimes \mathcal{A}$ defines a well-formed probability distribution over Σ^*.

Observe that $\mathcal{A}$ also defines a class of stochastic languages by the possible instantiations of T_j and F_j for each $\mathcal{M}_j \in \mathcal{A}$. The structural components of $\mathcal{A}$ are the structural components of each $\mathcal{M}_j \in \mathcal{A}$. By *structure* of $\mathcal{A}$, we mean its structural components. The structure of $\mathcal{A}$ defines a class of stochastic languages given by the possible instantiations of T_j and F_j satisfying Equation 1 for each $\mathcal{M}_j \in \mathcal{A}$.

If $\bigotimes \mathcal{A} = \mathcal{M}$ then the class of stochastic languages induced by the structure of $\mathcal{A}$ is a subset of the class of stochastic languages obtained with the structure of the PDFA $\mathcal{M}$. This is another way of saying that a factorized model may have fewer parameters and so the class of stochastic languages it represents can become smaller.

2.4 Statement of the Learning Problem

Let D be a finite sequence of $|D|$ i.i.d. drawn examples from a stochastic language L. It follows that the $P_L(D) = \prod_{w \in D} P_L(w)$.

Let $\mathcal{A} = \{\mathcal{M}_1 \dots \mathcal{M}_K\}$ be a set of PDFAs and let $C_{\mathcal{A}}$ denote the FRPD class of stochastic languages induced by the structure of $\mathcal{A}$. The *likelihood* of D w.r.t. $C_{\mathcal{A}}$ is determined by the parameters (the T_j and F_j functions for each $\mathcal{M}_j \in \mathcal{A}$). Let us group these parameters under the symbol Θ. Each Θ identifies some stochastic language $L_\Theta \in C_{\mathcal{A}}$. The *likelihood* of D w.r.t. $C_{\mathcal{A}}$ is defined as follows:

$$\mathsf{lhd}(D \mid \Theta) = \prod_{w \in D} P_{L_\Theta}(w).$$

The problem of finding a Maximum Likelihood Estimate (MLE) is to find those parameter values $\hat{\Theta}$ of $\mathcal{A}$ that maximize the likelihood of D w.r.t. $C_{\mathcal{A}}$. Formally,

$$\hat{\Theta} = \arg\max_{\Theta} \big(\mathsf{lhd}(D \mid \Theta) \big) \qquad (3)$$

where Θ under the $\arg\max$ ranges over all possible parameter values of $\mathcal{A}$.

When $|\mathcal{A}| = 1$ the problem has a known solution. As mentioned, a single PDFA $\mathcal{M}$ defines a class of stochastic languages given by possible parameter values of $\mathcal{M}$. In this case, it is well-known how to find $\hat{\Theta}$. Essentially, each transition probability $T(q, \sigma)$ equals the relative frequency that symbol σ is emitted at a state q (Vidal et al., 2005a,b). In this paper, we solve this problem when $|\mathcal{A}| > 1$.

3 Strictly k-Piecewise stochastic languages

In this section, we introduce the Strictly k-Piecewise stochastic languages, which serve as a running example of a FRPD class in the remainder of the paper.

Rogers et al. (2010) define and provide multiple characterizations of Strictly Piecewise (SP) languages. We review the most relevant ones for this paper here. SP languages are exactly those formal languages that are closed under subsequence.

$$\mathsf{SP} = \{L \subseteq \Sigma^* \mid \ \forall w, v \in \Sigma^* \\ (v \in L, w \sqsubseteq v \Rightarrow w \in L)\}$$

Rogers et al. (2010, p. 260) prove that every SP language L can be associated with a finite set of strings S such that L is the intersection of the complements of the shuffle ideals of S.

Theorem 1 $\forall L \in \text{SP}, \exists S \subseteq \Sigma^*, n \in \mathbb{N}$ *such that* $|S| < n$ *and* $L = \bigcap_{w \in S} \overline{SI(w)}$.

The SP languages are parameterized by a value $k \in \mathbb{N}$. This number corresponds to the length of the longest string in S. For each SP language L, if there is a set S whose longest string is equal to k, then L belongs to the SP_k class of languages.

If k is known a priori then the SP_k languages are both PAC-learnable and identifiable in the limit in polynomial time and data (Heinz, 2010b; Heinz et al., 2012).[1]

Theorem 1 allows one to construct concrete computational models for SP languages with DFA. For any nonempty string $w = \sigma_1 \ldots \sigma_n$, $SI(w) = L(\mathcal{M}_w)$ where $\mathcal{M}_w$ is defined as follows. The states are the prefixes of w, the start state is λ, and the final state is w. For all prefixes p of w and $\sigma \in \Sigma$, let $\delta(p, \sigma) = p\sigma$ whenever $p\sigma$ is a prefix of w and p otherwise. Figure 1 gives an examples of DFA for $\mathcal{M}_{abba}$.

The complement $\overline{SI(w)}$ is essentially obtained from $\mathcal{M}_w$ by removing its maximal state and making every state final. In other words, if $w = va$ then the $\overline{SI(w)}$ can be recognized by an automaton where the states are the prefixes of v, the start state is λ, and each state is a final state. For all prefixes p of v and $\sigma \in \Sigma$, $\delta(p, \sigma) = p\sigma$ whenever $p\sigma$ is a prefix of v. When $p\sigma$ is not a prefix of v and $\sigma \neq a$ then $\delta(p, \sigma) = p$. Finally, $\delta(v, a)$ is not defined. We denote such a DFA as $\mathcal{M}_{\overline{w}}$. Figure 2 shows the DFA $\mathcal{M}_{\overline{abba}}$ which recognizes the complement of $SI(abba)$. Both $\mathcal{M}_w$ and the DFA recognizing its complement are minimal.

It follows that for any $L \in \text{SP}$, one can construct a DFA recognizing L by taking the *product* of the complements of the shuffle ideals of the strings in S.

Note the size of $\mathcal{M}_1 \ldots \mathcal{M}_K$ is $\sum_{1 \leq i \leq K} \mathcal{M}_j$ whereas the size of $\mathcal{M} = \bigotimes_{1 \leq j \leq K} \mathcal{M}_j$ is in the worst case $\prod_{1 \leq j \leq K} \mathcal{M}_j$. Therefore, to decide whether a string w belongs to some SP language L, it may be preferable to run w on each $\mathcal{M}_j$ instead of on $\mathcal{M}$ to avoid the potentially large in-

crease in the state space. See Heinz and Rogers (2013) for additional discussion of this point.

Heinz and Rogers (2010) use the fact that SP languages are the intersection of the complements of shuffle ideals to define their stochastic counterpart. They define stochastic versions of $\mathcal{M}_{\overline{w}}$ (Figure 2), which they call *w-subsequence-distinguishing PDFA*.

Definition 2 (Subsequence-distinguishing PDFA) *Let* $w \in \Sigma^{k-1}$ *and* $w = \sigma_1 \cdots \sigma_{k-1}$. $\mathcal{M}_w = \langle Q, \Sigma, q_0, \delta, F, T \rangle$ *is a w-subsequence-distinguishing PDFA (w-SD-PDFA) iff F and T satisfy Equation 1 and* $\delta(u, \sigma) = u\sigma$ *whenever* $u\sigma \in \text{Pref}(w)$ *and u otherwise.*

Apart from the stochastic components T and F, the w-subsequence-distinguishing PDFA differs from $\mathcal{M}_{\overline{w}}$ in one key way. Suppose. $w = va$. Then $\delta(v, a) = v$ in the w-subsequence-distinguishing PDFA is not undefined as was the case with $\mathcal{M}_{\overline{w}}$. This transition exists and may have a nonzero probability.

A set $\mathcal{A}$ of PDFAs is *a k-set of SD-PDFAs* iff, for each $w \in \Sigma^{\leq k-1}$, it contains exactly one w-SD-PDFA. For example, let $\Sigma = \{a, b\}$ and consider the 2-set of SD-PDFAs shown in Figure 3. There are three SD-PDFAs in this set corresponding to $\mathcal{M}_\lambda, \mathcal{M}_a,$ and $\mathcal{M}_b$.

Heinz and Rogers (2010) define SP_k stochastic languages as a product of a k-set of SD-PDFAs. Specifically, the adapt the notion of co-emission probability (Vidal et al., 2005a). Heinz and Rogers (2010) actually use what they call the *positive co-emission product* which restricts the standard co-emission probability to particular circumstances.

In this work, we define SP stochastic languages with the standard definition of *co-emission probability* used to define products of PDFA as in Definition 1 (Vidal et al., 2005a).

Definition 3 (SP Stochastic Languages) *A probability distribution P over* Σ^* *is a SP stochastic language iff there exists a k-set of SD-PDFAs* $\mathcal{A}$, *whose co-emission product is* $\mathcal{M} = \bigotimes \mathcal{A}$, *such that for all* $w \in \Sigma^*$, *it is the case that* $P(w) = \mathcal{M}(w)$.

It follows immediately from this definition that the class of SP stochastic languages is a FRPD class. In this case, the parameters of such a distribution are the T and F values on each w-subsequence-distinguishing PDFA in the k-set. In the example in Figure 3, there are thus 15 parameters of the model, 10 of which are free. This is be-

---

[1] Also, SP languages suggest a different representation for strings (Rogers et al., 2013), which inform machine learning in other ways. The winning paper of the SPiCE competition (Balle et al., 2016), in which machine learning models competed to best predict the next symbol in a natural and artificial sequences was won by Shibata and Heinz (2016), who integrated SP-style representations into a neural network.

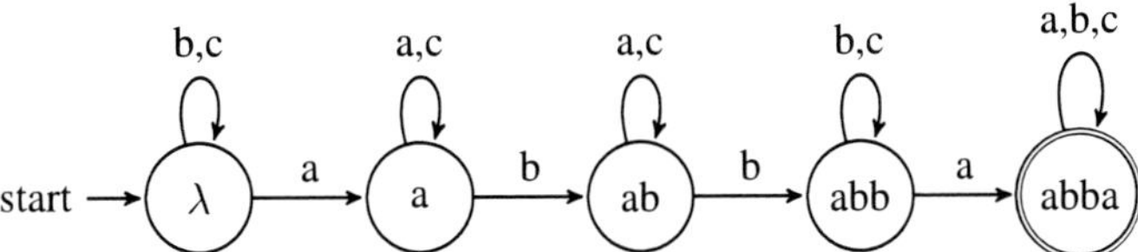

Figure 1: The DFA $\mathcal{M}_{abba}$ for $SI(abba)$ (left) with $\Sigma = \{a, b, c\}$.

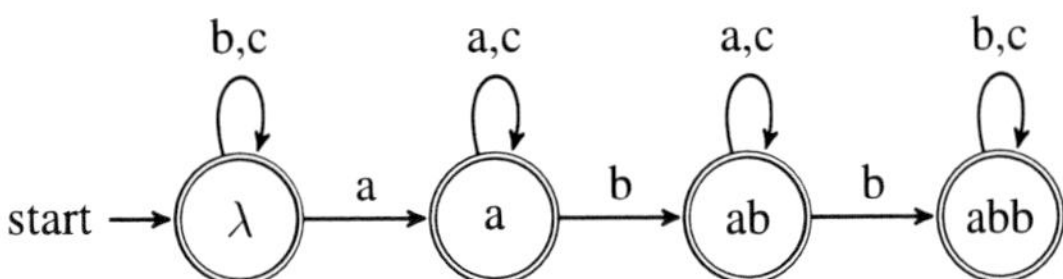

Figure 2: The DFA $\mathcal{M}_{\overline{abba}}$ for $\overline{SI(abba)}$ with $\Sigma = \{a, b, c\}$.

cause there are three actions associated with each state (a, b, and finality); there are five states; but since the probabilities must add to one only two parameters per state are free. More generally, a k-set of SD-PDFAs $\mathcal{A}$ has $|\Sigma| \cdot \sum_{j \in \mathcal{A}} |Q_j|$ free parameters.

4 Main Theorem for MLE of FRPD classes

We provide our main results here, using the 2-set of SD-PDFAs shown in Figure 3 as an illustrative example.

4.1 The Co-emission Probability Given a Prefix

It is useful to consider the co-emission probability of the symbol σ given the prefix $\sigma_1 \cdots \sigma_{i-1}$, which we denote $\mathrm{Coemit}(\sigma, i)$. It follows from Definitions 1 and 3 that this value is the normalized product of the path through $\bigotimes \mathcal{A}$ given by the prefix $\sigma_1 \cdots \sigma_{i-1}$.

Formally, let $M_1 = \langle Q_1, \Sigma, q_{01}, \delta_1, F_1, T_1 \rangle$, $\cdots$, $M_K = \langle Q_K, \Sigma, q_{0K}, \delta_K, F_K, T_K \rangle$ be exactly those PDFAs in $\mathcal{A}$. Suppose that $w = \sigma_1 \cdots \sigma_N$, where $\sigma_i \in \Sigma$ for all $1 \leq i \leq N$. Let $q(j, i)$ denote a state in Q_j that is reached after M_j reads the prefix $\sigma_1 \cdots \sigma_{i-1}$. If $i = 1$ then $q(j, i)$ represents the initial state of M_j. Then it follows from Definition 1 that the probability that a symbol σ is emitted after the product machine $\bigotimes_{1 \leq j \leq K} M_j$ reads the prefix $\sigma_1 \cdots \sigma_{i-1}$ is the following: $\mathrm{Coemit}(\sigma, i) =$

$$\frac{\prod_{j=1}^{K} T_j(q(j, i), \sigma)}{\sum_{\sigma' \in \Sigma} \left(\prod_{j=1}^{K} T_j(q(j, i), \sigma') \right) + \prod_{j=1}^{K} F_j(q(j, i))} \tag{4}$$

To simplify the notation and analysis, we assume that there is a end marker $\ltimes \in \Sigma$ which uniquely occurs at the end of words. This lets us replace $F_j(q)$ with $T_j(q, \ltimes)$. Then $\mathrm{Coemit}(\sigma, i)$ is simply written as

$$\mathrm{Coemit}(\sigma, i) = \frac{\prod_{j=1}^{K} T_j(q(j, i), \sigma)}{\sum_{\sigma' \in \Sigma} \prod_{j=1}^{K} T_j(q(j, i), \sigma').} \tag{5}$$

The probability that the machine $\bigotimes_{1 \leq j \leq K} M_j$ accepts w is obtained by taking the product of the co-emission probabilities for all i:

$$P(w\ltimes) = \prod_{i=1}^{N+1} \mathrm{Coemit}(\sigma_i, i), \tag{6}$$

where $\sigma_{N+1} = \ltimes$.

Since we are concerned with the co-emission probabilities, which is a ratio, it is noteworthy that in fact it does not matter if the sum $\sum_{\sigma' \in \Sigma} T_j(q, \sigma')$ is 1. The ratio $\mathrm{Coemit}(\sigma, i)$ and thus $P(w\ltimes)$ are invariant with respect to the scale of $T_j(q, \sigma')$ and the sum $\sum_{\sigma' \in \Sigma} T_j(q, \sigma')$. Writing this last value as $z(j, q)$, it can easily be confirmed by the fact that multiplying both the denominator and the numerator by $1/z(j, q)$ does not change the value of $\mathrm{Coemit}(\sigma, i)$ while normalizing $T_j(q, \cdot)$. Thus, we can relax the condition in Equation 1 when discussing co-emission probabilities. The only condition that needs to be satisfied with respect to the transitions is that $T_j(q, \sigma') \geq 0$ for all j, q, σ'. Note that relaxing this condition does not affect the number of free parameters. This is because the numerical values associated with the transitions, once normalized, will always sum to 1. In the following, we assume this relaxed condition.

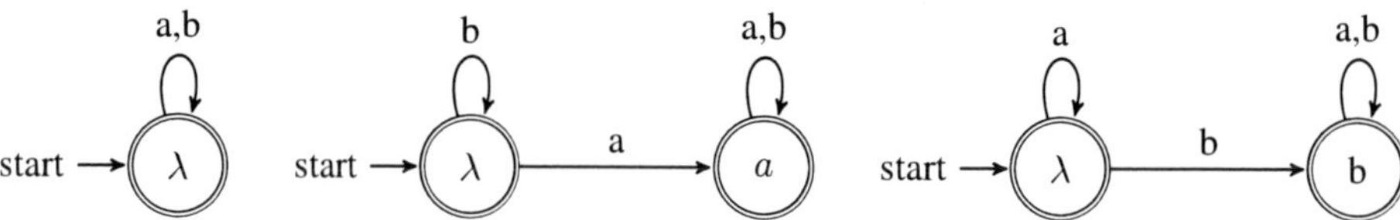

Figure 3: The 2-set of of SD-PDFAs with $\Sigma = \{a, b\}$.

4.2 Frequency and Empirical Mean of Co-emission Probability

Before describing the main theorem, we define two terms; *the frequency of an emission* and *the empirical mean of a co-emission probability*, which play important roles in estimating transition probabilities for product machines.

Definition 4 (Frequency of Emission) *For given w, we define the frequency of σ at $q \in Q_j$ as follows. Let*

- $m_w(M_j, q, \sigma) \in \mathbb{Z}^+$ *denotes how many times σ is emitted at the state q while the machine M_j emits w.*

- $n_w(M_j, q) \in \mathbb{Z}^+$ *denotes how many times the state q is visited while the machine M_j emits w.*

Then

$$\mathrm{freq}_w(\sigma | M_j, q) = \frac{m_w(M_j, q, \sigma)}{n_w(M_j, q)}, \qquad (7)$$

So $\mathrm{freq}_w(\sigma | M_j, q)$ represents the relative frequency that M_j emits σ at q during emission of w.

These concepts can be lifted to a sequence of strings D drawn i.i.d. from some stochastic language. Let

$$m_D(M_j, q, \sigma) = \sum_{w \in D} m_w(M_j, q, \sigma)$$

and

$$n_D(M_j, q) = \sum_{w \in D} n_w(M_j, q).$$

It follows that

$$\mathrm{freq}_D(\sigma | M_j, q) = \frac{m_D(M_j, q, \sigma)}{n_D(M_j, q)}.$$

So $\mathrm{freq}_D(\sigma | M_j, q)$ represents the relative frequency that M_j emits σ at q during emission of D.

As an example, consider the 2-set of PDFAs in Figure 3 and consider the sample data $D =$

$\langle abb\ltimes, aba\ltimes \rangle$. Figure 4 shows the paths of these strings through each SD-PDFA. Figure 5 shows some of the frequency computations.

If $K = 1$, i.e., the product machine consists of one PDFA then $\mathrm{freq}_w(\sigma | M_1, q)$ is the MLE of $T_1(q, \sigma)$ (Vidal et al., 2005a,b). Meanwhile, if $K \geq 2$, the probability of the emission, which equals the co-emission probability, fluctuates with states that other machines are currently at. Thus $\mathrm{freq}_w(\sigma | M_k, q)$, as a random variable, is not independent from other machines' states. This motivates the following definition.

Definition 5 (Empirical Mean) *Let*

$$\mathrm{sumCoemit}_w(\sigma, M_j, q) = \sum_{i \ s.t. \ q(j,i)=q} \mathrm{Coemit}(\sigma, i).$$

The empirical mean of a co-emission probability is defined as follows:

$$\overline{\mathrm{Coemit}}_w(\sigma | M_j, q) = \frac{\mathrm{sumCoemit}_w(\sigma, M_j, q)}{n_w(M_j, q)}, \qquad (8)$$

i.e., the sample average of the co-emission probability when $q \in Q_j$ is visited.

When a state in M_j is visited more than once while emitting w, it does not imply that some other state in M_h is also visited more than once. In other words, if there are positions $i \neq \ell$ such that $q(j, i) = q(j, \ell)$ then it does not have to follow that $q(h, i) = q(h, \ell)$ for another machine M_h. Thus, even when M_j and the value of $q(j, i)$ are fixed, $\mathrm{Coemit}(\sigma, i)$ fluctuates. The empirical mean is the average taken over such fluctuating co-emission probabilities.

4.3 Main Theorem and Convexity

Theorems 2 and 3 are our main results. We simplify the proofs by assuming that D consists of a single sentence. That is, in both theorems, we consider $D = \{w\ltimes\}$. We can do this without loss of generality because any finite sequence of strings D drawn i.i.d. from a stochastic language can be converted into a single sentence

$$\mathcal{M}_\lambda: \quad \lambda \xrightarrow{a} \lambda \xrightarrow{b} \lambda \xrightarrow{b} \lambda \xrightarrow{\ltimes} \qquad\qquad \lambda \xrightarrow{b} \lambda \xrightarrow{b} \lambda \xrightarrow{b} \lambda \xrightarrow{\ltimes}$$

$$\mathcal{M}_a: \quad \lambda \xrightarrow{a} a \xrightarrow{b} a \xrightarrow{b} a \xrightarrow{\ltimes} \qquad\qquad \lambda \xrightarrow{b} \lambda \xrightarrow{b} \lambda \xrightarrow{b} \lambda \xrightarrow{\ltimes}$$

$$\mathcal{M}_b: \quad \lambda \xrightarrow{a} \lambda \xrightarrow{b} b \xrightarrow{b} b \xrightarrow{\ltimes} \qquad\qquad \lambda \xrightarrow{b} b \xrightarrow{b} b \xrightarrow{b} b \xrightarrow{\ltimes}$$

Figure 4: The paths of $\{\mathrm{abb}\ltimes,\mathrm{bbb}\ltimes\}$ through the 2-set of of SD-PDFAs with $\Sigma = \{a,b\}$.

$$
\begin{aligned}
\mathrm{freq}_D(a|\mathcal{M}_\lambda,\lambda) &= 1/8 & \mathrm{freq}_D(a|\mathcal{M}_a,\lambda) &= 1/5 & \mathrm{freq}_D(a|\mathcal{M}_a,a) &= 0/3, \\
\mathrm{freq}_D(b|\mathcal{M}_\lambda,\lambda) &= 5/8 & \mathrm{freq}_D(b|\mathcal{M}_a,\lambda) &= 3/5 & \mathrm{freq}_D(b|\mathcal{M}_a,a) &= 2/3, \\
\mathrm{freq}_D(\ltimes|\mathcal{M}_\lambda,\lambda) &= 2/8 & \mathrm{freq}_D(\ltimes|\mathcal{M}_a,\lambda) &= 1/5 & \mathrm{freq}_D(\ltimes|\mathcal{M}_a,a) &= 1/3, \\[1em]
 & & \mathrm{freq}_D(a|\mathcal{M}_b,\lambda) &= 1/3 & \mathrm{freq}_D(a|\mathcal{M}_b,b) &= 3/5, \\
 & & \mathrm{freq}_D(b|\mathcal{M}_b,\lambda) &= 2/3 & \mathrm{freq}_D(b|\mathcal{M}_b,b) &= 0/5, \\
 & & \mathrm{freq}_D(\ltimes|\mathcal{M}_b,\lambda) &= 0/3 & \mathrm{freq}_D(\ltimes|\mathcal{M}_b,b) &= 2/5,
\end{aligned}
$$

Figure 5: Frequency computations with $D=\{\mathrm{abb}\ltimes,\mathrm{bbb}\ltimes\}$ and the 2-set of of SD-PDFAs in Figure 4.

without changing the probability of its production. To see why, we can adjust the transition function of each PDFA $\mathcal{M}_j$ so that $\delta_j(q,\ltimes) = q_{0j}$ for each $q \in Q_j$. In other words, once $\ltimes$ is emitted, the machines reset to their start states. Then for any $D = \{w_1\ltimes,\cdots,w_k\ltimes\}$, we have $P(D) = P(\mathrm{concat}(D))$ where $\mathrm{concat}(D) = w_1 \ltimes w_2 \ltimes \cdots \ltimes w_k\ltimes$. Thus, $w\ltimes$ in both theorems can be understood as $\mathrm{concat}(D)$.

Theorem 2 *Suppose that $P(w\ltimes)$ is defined as Equation 6 for a product machine $\bigotimes_{1\le j\le K} M_j$ and a word w. Then, $\partial P(w\ltimes)/\partial T_j = 0$ holds for all j if and only if the following equation is satisfied for all $1 \le j \le K$:*

$$\mathrm{freq}_w(\sigma|M_j,q) = \overline{\mathrm{Coemit}}_w(\sigma|M_j,q).$$

From Theorem 3, it will then follow that $T_1,\ldots T_K$ are the MLE.

Proof By taking the log of Eq. 6 , we have

$$
\begin{aligned}
\log P(w\ltimes) =& \sum_{i=1}^{N+1}\left(\sum_{j=1}^{K}\log T_j(q(j,i),\sigma_i)\right. \\
& \left. -\log\sum_{\sigma'\in\Sigma}\prod_{j=1}^{K}T_j(q(j,i),\sigma')\right) \\
=& \sum_{i=1}^{N+1}\sum_{j=1}^{K}\log T_j(q(j,i),\sigma_i) \\
& -\sum_{i=1}^{N+1}\log\sum_{\sigma'\in\Sigma}\prod_{j=1}^{K}T_j(q(j,i),\sigma').
\end{aligned}
$$

We differentiate this by a log emission probability $\log T_h(q,\sigma)$ for some $1 \le h \le K$. Let

$$A = \frac{\partial}{\partial\log T_h(q,\sigma)}\sum_{i=1}^{N+1}\sum_{j=1}^{K}\log T_j(q(j,i),\sigma_i),$$

and

$$B = \frac{\partial}{\partial\log T_h(q,\sigma)}\sum_{i=1}^{N+1}\log\sum_{\sigma'\in\Sigma}\prod_{j=1}^{K}T_j(q(j,i),\sigma').$$

Then

$$\frac{\partial}{\partial\log T_h(q,\sigma)}\log P(w\ltimes) = A - B.$$

First, we calculate A. Since

$$\frac{\partial T_j(q(j,i),\sigma_i)}{\partial\log T_h(q,\sigma)} = \begin{cases} 1 & \text{if } \langle M_h,q,\sigma\rangle \\ & = \langle M_j,q(j,i),\sigma_i\rangle, \\ 0 & \text{otherwise,} \end{cases}$$

we have

$$
\begin{aligned}
A &= \sum_{i=1}^{N+1}\sum_{j=1}^{K}\mathbf{I}\big[\langle M_h,q,\sigma\rangle = \langle M_j,q(j,i),\sigma_i\rangle\big] \\
&= \sum_{i=1}^{N+1}\mathbf{I}\big[\langle q,\sigma\rangle = \langle q(h,i),\sigma_i\rangle\big] \\
&= m_w(M_h,q,\sigma) \tag{9}
\end{aligned}
$$

where $\mathbf{I}[\cdot]$ denotes the indicator function and $m_w(M_h,q,\sigma)$ is defined as in Definition 4.

$$B = \frac{\partial}{\partial \log T_h(q,\sigma)} \sum_{i=1}^{N+1} \log \left(\sum_{a \in \Sigma} \prod_{j=1}^{K} T_j(q(j,i),a) \right)$$

$$= \sum_{i=1}^{N+1} \frac{\frac{\partial}{\partial \log T_h(q,\sigma)} \sum_{a \in \Sigma} \prod_{j=1}^{K} T_j(q(j,i),a)}{\sum_{a \in \Sigma} \prod_{j=1}^{K} T_j(q(j,i),a)}$$

$$= \sum_{i=1}^{N+1} \frac{\frac{\partial}{\partial \log T_h(q,\sigma)} \sum_{a \in \Sigma} \exp \left(\sum_{j=1}^{K} \log T_j(q(j,i),a) \right)}{\sum_{a \in \Sigma} \prod_{j=1}^{K} T_j(q(j,i),a)}$$

$$= \sum_{i=1}^{N+1} \frac{\sum_{a \in \Sigma} \left(\exp \left(\sum_{j=1}^{K} \log T_j(q(j,i),a) \right) \sum_{j=1}^{K} \frac{\partial \log T_h(q(j,i),a)}{\partial \log T_h(q,\sigma)} \right)}{\sum_{a \in \Sigma} \prod_{j=1}^{K} T_j(q(j,i),a)}$$

$$= \sum_{i=1}^{N+1} \frac{\sum_{a \in \Sigma} \left(\prod_{j=1}^{K} T_j(q(j,i),a) \sum_{j=1}^{K} \frac{\partial \log T_j(q(j,i),a)}{\partial \log T_h(q,\sigma)} \right)}{\sum_{a \in \Sigma} \prod_{j=1}^{K} T_j(q(j,i),a)}$$

$$= \sum_{i=1}^{N+1} \sum_{a \in \Sigma} \left(\frac{\prod_{j=1}^{K} T_j(q(j,i),a)}{\sum_{b \in \Sigma} \prod_{j=1}^{K} T_j(q(j,i),b)} \sum_{j=1}^{K} \frac{\partial \log T_j(q(j,i),a)}{\partial \log T_h(q,\sigma)} \right)$$

Figure 6: Initial calculation of B in the proof of Theorem 2.

Second, we calculate B as shown in Figure 6. There are two large terms in the large parentheses in the last line of the calculation of B in Figure 6. The first one is is the co-emission probability by Equation 5. Thus B =

$$\sum_{i=1}^{N+1} \sum_{a \in \Sigma} \sum_{j=1}^{K} \mathsf{Coemit}(a,i) \frac{\partial \log T_j(q(j,i),a)}{\partial \log T_h(q,\sigma)}.$$

Recall that

$$\frac{\partial \log T_j(q(j,i),a)}{\partial \log T_h(q,\sigma)}$$

equals

$$\mathbf{I}[\langle M_h,q,\sigma \rangle = \langle M_j,q(j,i),a \rangle].$$

This indicator function equals $\mathbf{I}[h = j]\mathbf{I}[q = q(j,i)]\mathbf{I}[\sigma = a]$. Abbreviating $\mathbf{I}[h = j]$ with $\mathbf{I}_1$, $\mathbf{I}[q = q(j,i)]$ with $\mathbf{I}_2$, and $\mathbf{I}[\sigma = a]$ with $\mathbf{I}_3$, we see that

$$\sum_{a \in \Sigma} \sum_{j=1}^{K} \mathsf{Coemit}(a,i)\,\mathbf{I}_1\mathbf{I}_2\mathbf{I}_3$$

$$= \sum_{j=1}^{K} \mathsf{Coemit}(\sigma,i)\,\mathbf{I}_1\mathbf{I}_2$$

$$= \mathsf{Coemit}(\sigma,i)\,\mathbf{I}[q = q(h,i)].$$

We conclude that

$$B = \sum_{i=1}^{N+1} \mathsf{Coemit}(\sigma,i)\mathbf{I}[q = q(h,i)]$$

$$= \sum_{i \text{ s.t. } q(h,i)=q} \mathsf{Coemit}(\sigma,i)$$

$$= \mathsf{sumCoemit}_w(\sigma,M_h,q). \tag{10}$$

By plugging our calculations of A (Eq. 9) and B (Eq. 10) into $A = B$ and dividing the both sides by $n_w(M_h,q)$, we obtain the result

$$\mathrm{freq}_w(\sigma|M_h,q) = \overline{\mathsf{Coemit}}_w(\sigma|M_h,q)$$

from the definitions of the relative frequency of an emission (Eq. 7) and the empirical mean of a co-emission probability (Eq. 8). This concludes the proof of Theorem 2. $\qquad\square$

Next we prove that maximizing $P(w)$ is a *convex optimization problem* to ensure that the solution is the maximum point.

Following Boyd and Vandenberghe (2004), A set of points C in $\mathbb{R}^n$ is *convex* if the line segment between any two points in C also lies in C. Formally, C is *convex* provided for any $x_1, x_2 \in C$ and any t with $0 \leq t \leq 1$, we have $tx_1 + (1 - t)x_2 \in C$. A function $f : \mathbb{R}^n \to \mathbb{R}$ is *convex* if

110

the domain of f is a convex set and if for all x, y in the domain of f, and t with $0 \leq t \leq 1$, we have $f(tx + (1-t)y) \leq tf(x) + (1-t)f(y)$. We say f is *concave* if $-f$ is convex.

Recall from section 2.4 that the likelihood of a sequence of data D to a stochastic language L belonging to a class with parameters Θ is $\mathsf{lhd}(D \mid \Theta) = \prod_{w \in D} P_L(w)$. The likelihood function is a function $f : \mathbb{R}^n \to \mathbb{R}$ where n is the number of parameters $|\Theta|$.

Let $\tau_{j,q,\sigma}$ denote $\log T_j(q, \sigma)$; i.e. the log of some parameter in Θ. There are $n = |\Sigma| \sum_{j=1}^{K} |Q_j|$ parameters in Θ since $\sigma \in \Sigma$, $1 \leq j \leq K$, and $q \in Q_j$. This τ can be thought of as a vector in $\mathbb{R}^n$.

The problem of maximizing $P(w \bowtie)$ is the same as minimizing $-\log P(w \bowtie)$ as a function of τ. We show that $\log P(w \bowtie)$ is concave with respect to $\log T_j(q, \sigma)$ (Theorem 3). If so, it is true that the solution shown in Theorem 2 is a global maximum.

Theorem 3 $\log P(w \bowtie)$ *is concave with respect to* $\tau \in \mathbb{R}^n$.

Proof By taking the log of Eq. 6 , we have $\log P(w \bowtie) =$

$$\sum_{i=1}^{N+1} \left(\sum_{j=1}^{K} \log T_j(q(j,i), \sigma_i) \right.$$

$$\left. - \log \sum_{a \in \Sigma} \prod_{j=1}^{K} T_j(q(j,i), a) \right).$$

Substituting in τ, it follows that $\log P(w \bowtie) =$

$$\sum_{i=1}^{N+1} \left(\sum_{j=1}^{K} \tau_{j,q(j,i),\sigma_i} - \log \sum_{a \in \Sigma} \prod_{j=1}^{K} \exp(\tau_{j,q(j,i),a}) \right) \cdot$$

Since

$$\prod_{j=1}^{K} \exp(\tau_{j,q(j,i),a}) = \exp \left(\sum_{k=1}^{K} \tau_{k,q(j,i),a} \right),$$

and by letting $g_a(\tau) = \sum_j \tau_{j,q(j,i),a}$, we obtain $\log P(w \bowtie) =$

$$\sum_{i=1}^{N+1} \left(g_{\sigma_i}(\tau) - \log \sum_{a \in \Sigma} \exp(g_a(\tau)) \right).$$

Generally speaking, a composition $f(x) = h(g_1(x), \cdots, g_k(x))$ obeys the following rule: f is convex if h is convex, h is non-decreasing in each argument, and g_i is convex (see vector composition in Boyd and Vandenberghe, 2004, section 3.2.4)). Furthermore, it is known that $\log \sum \exp(\cdot)$ is convex (see section 3.1.5), and $\log \sum \exp(\cdot)$ is non-decreasing in each argument since both $\exp(\cdot)$ and $\log(\cdot)$ are non-decreasing. In addition, $g_a(\cdot)$ is both convex and concave since every linear function is so from the definition (see section 3.1.1). Thus, $\log \sum_a \exp(g_a(\cdot))$ is convex, and $-\log \sum_a \exp(g_a(\cdot))$ is concave.

Finally, from the fact that non-negative weighted sum preserves convexity and concavity (Boyd and Vandenberghe, 2004, section 3.2.1), $\log P(w \bowtie)$ is concave. $\qquad \square$

It follows that the negative log of $P(w \bowtie)$ is convex.

It is noteworthy to point out that establishing concavity does not mean the solution is unique. In fact, the solutions can be a set of points. An example FRPD class illustrating this is one which contains two PDFA $\mathcal{M}_1$ and $\mathcal{M}_2$ with the same structure. For example suppose each had exactly one state with self-loop transitions for every symbol in Σ. The co-emission product $\mathcal{M}_1 \otimes \mathcal{M}_2$ does not uniquely factorize though the above theorem establishes its convexity.

Of course it is also of interest to know when the solution is unique. In this case, we have to show the negative log probability is strictly convex except for multiplying the emission probability by a constant. We leave this as an area of future research.

5   Optimization and Time Complexity

In this section, we discuss the time complexity and also how to optimize. From the proof of Theorem 2, we have the following fact immediately.

Corollary 1 *The update equation for maximization of* $\log P(w \bowtie)$ *is represented as:* $\log T_j(q, \sigma) :=$

$$\log T_j(q, \sigma) + \eta \left(\mathrm{freq}_w(\sigma | M_j, q) - \overline{\mathrm{Coemit}}_w(\sigma | M_j, q) \right) \qquad (11)$$

if the simplest gradient method is applied, and where η *is the step size. The time complexity for each update is* $O(NK|\Sigma|)$.

The time complexity for $\mathrm{freq}_w(\sigma | M_j, q)$ and $\overline{\mathrm{Coemit}}_w(\sigma | M_j, q)$ are shown in Lemma 1 and Lemma 2. The time complexity for

$\overline{\mathsf{Coemit}}_w(\sigma|M_j, q)$ is a little higher than that of $\mathsf{freq}_w(\sigma|M_j, q)$.

Lemma 1 *For all M_j and $q \in Q_j$, $\mathsf{freq}_w(\sigma|M_j, q)$ are computed in the time $O(NK)$.*

Proof We trace all machines while they are emitting $\sigma_1, \cdots, \sigma_N$. Suppose that machines are at $q(1, i), \cdots, q(K, i)$ after $\sigma_1, \cdots, \sigma_{i-1}$ are emitted sequentially. For each step i, for all machines M_j, we have to update the counting for the pair of $q(k, i)$ and σ_i, in order to calculate $m_w(M_j, q, \sigma)$. So the computational cost for each step i is $O(K)$. $\qquad\qquad\square$

Lemma 2 *For all M_j and $q \in Q_j$, $\overline{\mathsf{Coemit}}_w(\sigma|M_j, q)$ are computed in the time $O(NK|\Sigma|)$.*

Proof We trace all machines while they are emitting $\sigma_1, \cdots, \sigma_N$. Suppose that machines are at $q(1, i), \cdots, q(K, i)$ after $\sigma_1, \cdots, \sigma_{i-1}$ are emitted sequentially. The critical part is calculating $\mathsf{sumCoemit}(\sigma)_{\langle M_j, q \rangle}(w)$. For each step i, we have to update emission probabilities for all pairs of M_j and $\sigma \in \Sigma$. This update is in the time $O(K|\Sigma|)$. Thus, the time complexity for calculating $\mathsf{sumCoemit}_w(\sigma, M_j, q)$ is $O(NK|\Sigma|)$. $\square$

6 Conclusion

The negative log likelihood function associated with a FRPD class C is convex, and it is possible to efficiently find a MLE of any sequences of data generated i.i.d. with respect to C. Essentially, the parameters of the model are found by running the corpus through each of the individual factor PDFAs and calculating the relative frequencies. While this was the approach adopted by Heinz and Rogers (2010) for SP stochastic languages, we have generalized it to sets of finitely many PDFAs.

There are several directions for future research, both theoretical and applied. On the theoretical side, one clear avenue is to better understand these results in terms of probabilistic graphical models (Koller and Friedman, 2009). As a reviewer pointed out, the application of those methods to formal language theory and grammatical inference (de la Higuera, 2010) appears fruitful.

On the applied side, there are several different opportunities. One area of interest is language modeling. The results here permit a modular approach to constructing language models, where certain primitive factors are included or excluded. For example, we expect that language models which incorporate both n-gram models (Jurafsky and Martin, 2008) (which cannot describe long-distance dependencies) and SP stochastic languages (which can describe some kinds of long-distance dependencies) will have lower perplexity, a hypothesis under current investigation. More generally, researchers can use aspects of the sub-regular hierarchies of languages (Thomas, 1997; Rogers et al., 2013) to identify a range of 'primitive factors' whose DFA models can form the basis of various FRPD classes.

Finally, we are also interested in extending these results to weighted deterministic automata for computing regular relations (Beros and de la Higuera, 2016) or elements of other monoids (Gerdjikov, 2018).

Acknowledgments

We would like to thank two anonymous reviewers for helpful comments, and another anonymous reviewer in particular for making clear the scope of this work, which resulted in a significant revisions to our original submission. This work was supported by NIH grant #R01HD87133-01 to JH and JSPS KAKENHI grant #JP18K11449 to CS.

References

Borja Balle, Rémi Eyraud, Franco M. Luque, Ariadna Quattoni, and Sicco Verwer. 2016. Results of the sequence prediction challenge (SPiCe): a competition on learning the next symbol in a sequence. In *Proceedings of The 13th International Conference on Grammatical Inference*, volume 57 of *JMLR: Workshop and Conference Proceedings*, pages 132–136.

Achilles Beros and Colin de la Higuera. 2016. A canonical semi-deterministic transducer. *Fundamenta Informaticae*, 146(4):431–459.

Christopher M. Bishop. 2006. *Pattern Recognition and Machine Learning*. Information Science and Statistics. Springer.

S. Boyd and L. Vandenberghe. 2004. *Convex optimization*. Cambridge.

Stefan Gerdjikov. 2018. A general class of monoids supporting canonisation and minimisation of (sub)sequential transducers. In *Language and Automata Theory and Applications - 12th International Conference, LATA 2018, Ramat Gan, Israel, April 9-11, 2018, Proceedings*, pages 143–155.

J. Heinz and J. Rogers. 2010. Estimating Strictly Piecewise Distributions. *Proceedings of the 48th Annual Meeting of the Association for Computational Linguistics*, pages 886–896.

Jeffrey Heinz. 2010a. Learning long-distance phonotactics. *Linguistic Inquiry*, 41(4):623–661.

Jeffrey Heinz. 2010b. String extension learning. In *Proceedings of the 48th Annual Meeting of the Association for Computational Linguistics*, pages 897–906, Uppsala, Sweden. Association for Computational Linguistics.

Jeffrey Heinz. 2014. Culminativity times harmony equals unbounded stress. In Harry van der Hulst, editor, *Word Stress: Theoretical and Typological Issues*, chapter 8. Cambridge University Press, Cambridge, UK.

Jeffrey Heinz. 2018. The computational nature of phonological generalizations. In Larry Hyman and Frans Plank, editors, *Phonological Typology*, Phonetics and Phonology, chapter 5, pages 126–195. De Gruyter Mouton.

Jeffrey Heinz, Colin de la Higuera, and Menno van Zaanen. 2015. *Grammatical Inference for Computational Linguistics*. Synthesis Lectures on Human Language Technologies. Morgan and Claypool.

Jeffrey Heinz, Anna Kasprzik, and Timo Kötzing. 2012. Learning with lattice-structured hypothesis spaces. *Theoretical Computer Science*, 457:111–127.

Jeffrey Heinz and James Rogers. 2013. Learning subregular classes of languages with factored deterministic automata. In *Proceedings of the 13th Meeting on the Mathematics of Language (MoL 13)*, pages 64–71, Sofia, Bulgaria. Association for Computational Linguistics.

Colin de la Higuera. 2010. *Grammatical Inference: Learning Automata and Grammars*. Cambridge University Press.

John Hopcroft, Rajeev Motwani, and Jeffrey Ullman. 2001. *Introduction to Automata Theory, Languages, and Computation*. Boston, MA: Addison-Wesley.

Daniel Jurafsky and James Martin. 2008. *Speech and Language Processing: An Introduction to Natural Language Processing, Speech Recognition, and Computational Linguistics*, 2nd edition. Prentice-Hall, Upper Saddle River, NJ.

Daphne Koller and Nir Friedman. 2009. *Probabilistic Graphical Models: Principles and Techniques*. MIT Press.

Robert Malouf. 2002. A comparison of algorithms for maximum entropy parameter estimation. In *Proceedings of the 6th Conference on Natural Language Learning - Volume 20*, COLING-02, pages 1–7. Association for Computational Linguistics.

James Rogers, Jeffrey Heinz, Gil Bailey, Matt Edlefsen, Molly Visscher, David Wellcome, and Sean Wibel. 2010. On languages piecewise testable in the strict sense. In *The Mathematics of Language*, volume 6149 of *Lecture Notes in Artifical Intelligence*, pages 255–265. Springer.

James Rogers, Jeffrey Heinz, Margaret Fero, Jeremy Hurst, Dakotah Lambert, and Sean Wibel. 2013. Cognitive and sub-regular complexity. In *Formal Grammar*, volume 8036 of *Lecture Notes in Computer Science*, pages 90–108. Springer.

Chihiro Shibata and Jeffrey Heinz. 2016. Predicting sequential data with lstms augmented with strictly 2-piecewise input vectors. In *Proceedings of The 13th International Conference on Grammatical Inference*, volume 57 of *JMLR: Workshop and Conference Proceedings*, pages 137–142.

Wolfgang Thomas. 1997. Languages, automata, and logic. In *Handbook of Formal Languages*, volume 3, chapter 7. Springer.

Enrique Vidal, Franck Thollard, Colin de la Higuera, Francisco Casacuberta, and Rafael C. Carrasco. 2005a. Probabilistic finite-state machines-part I. *IEEE Transactions on Pattern Analysis and Machine Intelligence*, 27(7):1013–1025.

Enrique Vidal, Frank Thollard, Colin de la Higuera, Francisco Casacuberta, and Rafael C. Carrasco. 2005b. Probabilistic finite-state machines-part II. *IEEE Transactions on Pattern Analysis and Machine Intelligence*, 27(7):1026–1039.

Sentence Length

Gábor Borbély   **András Kornai**
{borbely,kornai}@math.bme.hu
Department of Algebra, Budapest University of Technology and Economics

Abstract

The distribution of sentence length in ordinary language is not well captured by the existing models. Here we survey previous models of sentence length and present our random walk model that offers both a better fit with the data and a better understanding of the distribution. We develop a generalization of KL divergence, discuss measuring the noise inherent in a corpus, and present a hyperparameter-free Bayesian model comparison method that has strong conceptual ties to Minimal Description Length modeling. The models we obtain require only a few dozen bits, orders of magnitude less than the naive nonparametric MDL models would.

1 Introduction

Traditionally, statistical properties of sentence length distribution were investigated with the goal of settling disputed authorship (Mendenhall, 1887; Yule, 1939). Simple models, such as a "monkeys and typewriters" Bernoulli process (Miller, 1957) do not fit the data well, and this problem is inherited from n-gram Markov to n-gram Hidden Markov models, such as found in standard language modeling tools like SRILM (Stolcke et al., 2011). Today, length modeling is used more often as a downstream task to probe the properties of sentence vectors (Adi et al., 2017; Conneau et al., 2018), but the problem is highly relevant in other settings as well, in particular for the current generation of LSTM/GRU-based language models that generally use an ad hoc cutoff mechanism to regulate sentence length. The first modern study, interested in the entire shape of the sentence-length distribution, is Sichel (1974), who briefly summarizes the earlier proposals, in particular negative binomial (Yule, 1944), and lognormal (Williams, 1944), being rather critical of the latter:

The lognormal model suggested by Williams and used by Wake must be rejected on several grounds: In the first place the number of words in a sentence constitutes a discrete variable whereas the lognormal distribution is continuous. Wake (1957) has pointed out that most observed log-sentence-length distributions display upper tails which tend towards zero much faster than the corresponding normal distribution. This is also evident in most of the cumulative percentage frequency distributions of sentence-lengths plotted on log-probability paper by Williams (1970). The sweep of the curves drawn through the plotted observations is concave upwards which means that we deal with sub-lognormal populations. In other words, most of the observed sentence-length distributions, after logarithmic transformation, are negatively skew. Finally, a mathematical distribution model which cannot fit real data –as shown up by the conventional χ^2 test– cannot claim serious attention. (Sichel, 1974, p. 26)

Sichel's own model is a mixture of Poisson distributions given as

$$\phi(r) = \frac{\sqrt{1-\theta}^{\,\gamma}}{K_\gamma(\alpha\sqrt{1-\theta})} \frac{(\alpha\theta/2)^r}{r!} K_{r+\gamma}(\alpha) \quad (1)$$

where K_γ is the modified Bessel function of the second kind of order γ. As Sichel notes, "a number of known discrete distribution functions such as the Poisson, negative binomial, geometric, Fisher's logarithmic series in its original and modified forms, Yule, Good, Waring and Riemann distributions are special or limiting forms of (1)".

While Sichel's own proposal certainly cannot be faulted on the grounds enumerated above, it still leaves something to be desired, in that the parameters α, γ, θ are not at all transparent, and the model lacks a clear genesis. In Section 2 of this article we present our own model aimed at remedying these defects and in Section 3 we analyze its properties. Our results are presented is Section 4. The relation between the sentence length model and grammatical theory is discussed in the concluding Section 5.

2 The random walk model

In the following Section we introduce our model of random walk(s). The predicted sentence length is basically the return time of these stochastic processes, i.e. the probability of a given length is the probability of the appropriate return time.

Let X_k be a random walk on $\mathbb{Z}$ and $X_k(t)$ the position of the walk at time t. Let $X_k(0) = k$ be the initial condition. The walk is given by the following parameters:

$$X_k(t+1) - X_k(t) = \begin{cases} -1 & \text{with probability } p_{-1} \\ 0 & \text{with probability } p_0 \\ 1 & \text{with probability } p_1 \\ 2 & \text{with probability } p_2 \end{cases} \tag{2}$$

The random walk is the sum of these independent steps. (2) is a simple model of valency (dependency) tracking: at any given point we may introduce, with probability p_2, some word with two open valences (e.g. a transitive verb), with probability p_1 one that brings one new valence (e.g. an intransitive verb or an adjective), with probability p_0 one that doesn't alter the count of open valencies (e.g. an adverbial), and with probability p_{-1} one that fills an open valency, e.g. a proper noun. For ease of presentation here we cut off at 2, making no provisions for ditransitives and higher arity verbs, but in actual numerical work (Section 3) we will relax this assumption. We also cut off at -1, making no provision for those cases where a single word can fill more than one valency, as in Latin *accusativus cum infinitivo* or (arguably) English equi. We discuss these cutoffs further in Sections 3.1 and 5. The return time is defined as

$$\tau_k = \min_{t \geq 0}\{t : X_k(t) = 0\} \tag{3}$$

In particular, τ_1 is the time needed to go from $1 \rightarrow$

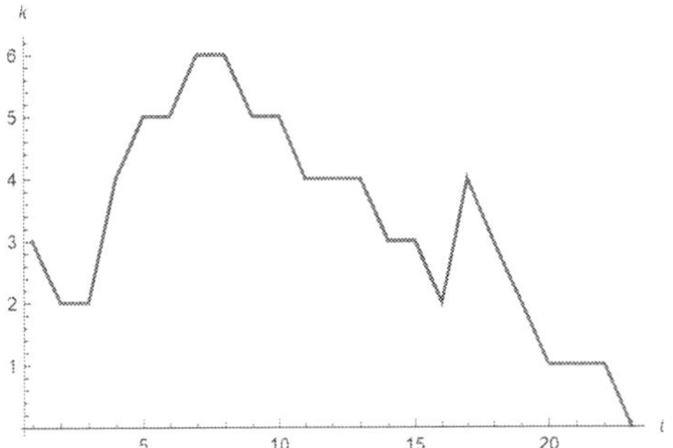

Figure 1: Sentence length is modeled as the return time of a random walk.

0. We will calculate the probability-generating function to find the probabilities.

$$f(x) := \mathbb{E}\left(x^{\tau_1}\right) \tag{4}$$

The generating function of τ_k easily follows from τ_1, since τ_k is the sum of k independent copies of τ_1, so the generating function of τ_k is simply $f(x)^k$.

In order to calculate $f(x)$, we condition on the first step:

$$\begin{aligned} f(x) = p_{-1} \cdot x + & \qquad \text{finishing in one step} \\ p_0 \cdot x \cdot f(x) + & \qquad \text{wait } \tau_1 \text{ again} \\ p_1 \cdot x \cdot f(x)^2 + & \qquad \text{wait } \tau_1 \text{ two times} \\ p_2 \cdot x \cdot f(x)^3 & \qquad \text{wait } \tau_1 \text{ three times} \end{aligned} \tag{5}$$

Therefore, $f(x)$ is the solution of the following equation (solved for f, x is a parameter):

$$p_{-1} \cdot x + (p_0 \cdot x - 1) \cdot f + p_1 \cdot x \cdot f^2 + p_2 \cdot x \cdot f^3 = 0 \tag{6}$$

This can be solved with Cardano's formula. The probabilities are given by

$$\mathbb{P}(\tau_k = i) = [x^i] f(x)^k \triangleq \frac{1}{i!} \frac{\partial^i}{\partial x^i} f(x)^k \bigg|_{x=0} \tag{7}$$

(Here and in what follows, $[x^i]$ refers to the coefficient of x^n in the expansion of the function to the right of it.) For given parameters p_{-1}, p_0, p_1, p_2 and k, and a given i, one can evaluate these probabilities numerically, but we need a bit more analytical form. Let us define the following.

$$F(u) = p_{-1} + p_0 \cdot u + p_1 \cdot u^2 + p_2 u^3 \tag{8}$$

$$g(f) = \frac{f}{F(f)} \tag{9}$$

With these functions Equation 6 becomes $x = g(f(x))$, meaning that we are looking for the inverse function of g. One can see that $g(0) = 0$ and $g'(0) = 1/p_{-1} \neq 0$, therefore we can apply the Lagrange inversion theorem. Calculations detailed in the Appendix yield the following formula.

$$\mathbb{P}(\tau_k = i) = \frac{k}{i}[u^{i-k}]F^i(u) \qquad (10)$$

Since F is a polynomial, one can calculate its powers by polynomial multiplication and get $\mathbb{P}(\tau_k = i)$ by looking up the appropriate coefficient. Here k is an integer (discrete) model parameter and p_{-1}, p_0, p_1, p_2 are real (continuous) numbers. This makes the above mentioned probabilities *differentiable* in the continuous parameters.

We call the parameter k, the starting point of the random walk, the total valency. Note that $\tau_k \geq k$ with probability 1, therefore one cannot model the sentences shorter then k. To overcome this obstacle, we introduce the mixture model that consists of several models with various k values and coefficients for convex linear combination.

$$\mathbb{P}_{k_1,\alpha_1,k_2,\alpha_2,\ldots k_m,\alpha_m}(\tau = i) = \sum_{j=1}^{m} \alpha_j \cdot \mathbb{P}(\tau_{k_j} = i)$$
$$(11)$$

where the parameters α_j are mixture coefficients; positive and sum up to 1, see Figure 2. Also every term in the mixture have different p_{-1}, p_0, p_1 and p_2 values (all positive and sum up to one). In this way, we can model the sentences with length at least $\min_j k_j$.

k_1	α_1	p_{-1}^1	p_0^1	p_1^1	p_2^1
k_2	α_2	p_{-1}^2	p_0^2	p_1^2	p_2^2
$\vdots$	$\vdots$		$\vdots$		
k_m	α_m	p_{-1}^m	p_0^m	p_1^m	p_2^m

Figure 2: Model parameters. The framed parameters are real, positive numbers and should sum up to 1.

It is easy to generalize our model to allow higher upward steps, i.e. p_3 for ditransitives or even higher steps for higher arity relations. The only technical constraint is that p_{-1} and p_0 should be positive, and no lower steps are allowed (no p_{-2}). This is also a reasonable assumption if a word can only fulfill one role in a sentence, a matter we return to in Section 5. Altogether, the number of upward steps is called *order* and it is another hyper-parameter in our model.

Theoretically, there is no obstacle to have different number of p values to different k values. The model can be a heterogeneous mixture of random walks, where the individual processes can have different upward steps. But we did not investigate that possibility.

3 Model analysis

Here we introduce and analyze the experimental setup that we will use in Section 4 to fit our model to various datasets. The raw data is a set of positive integers, the sentence lengths, and their corresponding weights (absolute frequencies) $\{n_x\}_{x \in X}$. We call $n := \sum_{x \in X} n_x$ the *size* and X the *support* of the data. Since the model is differentiable in the continuous parameters (including the mixing coefficients), the direct approach would be to perform gradient descent on the dissimilarity as an objective function to find the parameters. With fixed valency parameters k_j this is a constrained optimization task $\text{dist}(\mathbb{P}_{\text{sample}}, \mathbb{P}_{\text{modeled}}) \to \min$.

In some cases, especially for smaller datasets, we might find it expedient to bin the data, for example (Adi et al., 2017) use bins (5-8), (9-12), (13-16), (17-20), (21-25), (26-29), (30-33), and (34-70). On empirical data (for English we will use the BNC[1] and the UMBC Webbase[2] and for other languages the SZTAKI corpus[3]) this particular binning leaves a lot to be desired. We discuss this matter in subsection 3.1, together with the choice of dissimilarity (figure of merit). An important consideration is that a high number of mixture components fit the data better but have more model parameters – this is discussed in subsection 3.2.

3.1 Length extremes

Short utterances are common both in spoken corpora and in written materials, especially in dialog intended to sound natural (see 2nd and 5th columns of Table 1). As is well known, people don't speak in complete sentences, and a great deal of the short material is the result of sluicing, zero anaphora, and similar cross-sentence ellipsis

[1] http://www.natcorp.ox.ac.uk
[2] https://ebiquity.umbc.edu/resource/html/id/351
[3] http://hlt.sztaki.hu/resources/webcorpora.html

phenomena (Merchant, 2001), with complete sentences such as imperatives like *Help!* comprising only a small portion of the data. In nonfiction, short strings are encountered overwhelmingly in titles, subtitles, and itemized lists, material that is hard to separate from actual sentences. Here we go around the problem by permitting in the mixture components with low total valency (small k at the start of the random walk).

dataset	< 5	> 100	dataset	< 5	> 100
BNC-A	7.2%	0.1%	Dutch	17.4%	1.1%
BNC-B	9.6%	0.1%	Finnish	14.1%	0.7%
BNC-C	8.8%	0.1%	Indonesian	11.3%	2.0%
BNC-D	25.9%	1.4%	Lithuanian	25.2%	1.1%
BNC-E	8.7%	0.1%	Bokmål	14.4%	1.1%
BNC-F	12.1%	0.2%	Nynorsk	8.7%	0.4%
BNC-G	11.2%	0.1%	Polish	23.3%	1.9%
BNC-H	14.5%	0.2%	Portuguese	22.7%	2.5%
BNC-J	15.2%	0.5%	Romanian	8.2%	3.1%
BNC-K	29.9%	0.2%	Serbian.sh	15.3%	1.9%
UMBC	3.7%	0.2%	Serbian.sr	33.7%	9.0%
Catalan	15.7%	2.8%	Slovak	12.4%	1.9%
Croatian	16.7%	2.1%	Spanish	14.7%	3.2%
Czech	13.7%	1.3%	Swedish	24.6%	0.8%
Danish	20.8%	1.1%			

Table 1: Distribution of short and long sentences

Especially on the long end (see columns 3 and 6 of Table 1) data becomes so sparse that some kind of binning is called for. Since the eight bins used by (Adi et al., 2017) actually ignore the very low (1-4) and very high (71+) ranges of the data, we will use ordinary deciles, setting the ten bins as the data dictates. In this regard, it is worth noting that in the 18 non-English corpora used in this study the low bin neglected by (Adi et al., 2017) contains on the average 17.4% of the data (variance 6.3%, low 8.1% on Romanian, high 33.7% on Serbian_sr) whereas on the high end the problem is much less severe: for example in UMBC 1.0%, and in the BNC only 0.8% would be ignored.

To cover 99.9% we need to consider only sentences up to a few hundred words (see column 5 of Table 2), and in the current study we applied a cutoff of 1,000 to be above 99.9% coverage in all cases while keeping compute time manageable. The last column of Table 2 shows the length of the longest sentence in each of the subcorpora considered. The original binning (cutoff at 71) would have resulted in coverage 95.7% on the average (variance 3.1%, low 84.9% Serbian_sr, high 98.8% for Nynorsk).

The prevailing tokenization convention, where punctuation is counted as equivalent to a full word,

dataset		number of sentences	tolerance (in nats)	mean	99.9% sentence length	max
BNC-2.0 (English)	BNC-A	753442	9.847e-4	20.967	97	555
	BNC-B	362003	7.741e-3	20.650	96	365
	BNC-C	955486	9.494e-3	20.524	102	491
	BNC-D	6138	8.510e-2	16.366	228	466
	BNC-E	337370	5.000e-3	22.219	106	763
	BNC-F	527758	2.630e-2	19.351	130	2208
	BNC-G	478860	9.199e-3	18.753	106	435
	BNC-H	1185549	3.385e-2	18.841	118	950
	BNC-J	359352	7.940e-2	18.666	156	1100
	BNC-K	1086242	2.134e-1	12.784	116	918
	UMBC	136630947	2.442e-3	24.434	116	3052
SZTAKI corpus	Catalan	23927377	1.751e-3	27.496	384	5279
	Croatian	62196524	5.616e-3	23.975	369	8598
	Czech	30382696	5.147e-3	20.139	285	6081
	Danish	26687240	7.557e-3	18.593	296	16425
	Dutch	103958658	2.408e-3	19.135	296	16128
	Finnish	58104101	1.946e-3	15.538	237	5552
	Indonesian	13095607	1.231e-2	23.675	343	22762
	Lithuanian	81826291	1.184e-3	17.170	294	21857
	Bokmål	84375397	3.564e-3	19.199	281	14032
	Nynorsk	1393312	3.946e-3	18.836	175	1591
	Polish	72983880	8.508e-3	19.549	396	24353
	Portuguese	37953728	4.973e-2	25.365	448	9614
	Romanian	36211510	2.338e-2	29.466	473	54434
	Serbian.sh	35606837	4.531e-3	23.744	332	6800
	Serbian.sr	2023815	7.189e-3	37.736	862	6800
	Slovak	39633566	2.572e-3	21.759	402	24571
	Spanish	47673229	8.365e-4	29.305	471	29183
	Swedish	54218846	2.526e-3	16.468	315	8127

Table 2: Sentence length datasets. For tolerance see subsection 3.2

has an effect on the distribution, more perceptible at the low end. Besides this (and more subtle issues of tokenization, such as the treatment of hyphenation or of multiple punctuation) perhaps the most important factor influencing sentence length is morphological complexity, since in highly agglutinating languages a single word is sufficient for what would require a multiword sentence in English, as in Hungarian *elvihetlek* 'I can give you a ride'.

Since the number of datapoints is high, ranging from 1.3M (Nynorsk) to 136.6M (UMBC), the conventional χ^2 test does not provide a good figure of merit on the original data (no fit is ever significant, especially as there is a lot of variation at the high end where only few lengths are extant), nor on the binned data, where every fit is highly significant.

A better choice is the Kullback–Leibler divergence, but this still suffers from problems when the supports of the distributions do not coincide. In our case we have this problem both at the low end, where the model predicts $\mathbb{P}(\tau = i) = 0$ for $i < k$, and at the high end, where we predict positive (albeit astronomically small) probabilities of arbitrarily long sentences. To remedy this defect, we define generalized KL divergence, gKL, as follows.

Definition 3.1 (Motivated by Theorem A.2.). *Let $\mathbb{P}$ and $\mathbb{Q}$ be probability measures over the same measurable space (X, Σ) that are both absolutely continuous with respect to a third measure dx, and let λ be $\mathbb{P}(\text{supp}(\mathbb{P}) \cap \text{supp}(\mathbb{Q}))$. Then*

$$gKL(\mathbb{P}, \mathbb{Q}) := -\lambda \cdot \ln \lambda +$$
$$\int\limits_{\text{supp}(\mathbb{P}) \cap \text{supp}(\mathbb{Q})} \mathbb{P}(x) \cdot \ln \frac{\mathbb{P}(x)}{\mathbb{Q}(x)} \, dx \tag{12}$$

Clearly, gKL reduces to the usual KL divergence if the support of the distributions coincide. The high end of the distribution could be ignored, at least for English, at the price of losing less than 0.1% of the data, but ignoring the short sentences, 14.4% of the BNC, is hard to countenance. As a practical matter this means we needed to bring in mixture components with total valency $k < 4$, and these each bring 4 parameters (the mixture weight and 3 p_i values) in tow. Obviously, the more components we use, the better the fit will be, so we need to control the trade-off between these. In subsection 3.2 we introduce a method derived from Bayesian model comparison (MacKay, 2003) that will remedy the zero modeled probabilities and answer the model complexity trade-off.

3.2 Bayesian model comparison

If a dataset D has support X, with $n_x > 0$ being the number that length x occurred, the data size is $|D| = \sum_{x \in X} n_x$ and the observed probabilities are $p_x := \frac{n_x}{|D|}$. Let $\mathcal{H}_i \subseteq \mathbb{R}^d$ be i^{th} model in some list of models. Each model is represented by a parameter vector $\mathbf{w}_i \in \mathcal{H}_i$ in the parameter space, and supp $\mathcal{H}_i = \{x \mid \mathbb{P}(x \mid \mathcal{H}_i) > 0\}$ is not necessarily equal to X. Clearly, different $\mathcal{H}_i$ may have different support, but a given model has the same support for every $\mathbf{w}_i$. Model predictions are given by $\mathbb{Q}_{\mathbf{w}_i}(x) := \mathbb{P}(x \mid \mathbf{w}_i, \mathcal{H}_i)$, and the **evidence** the ith model has is

$$\mathbb{P}(\mathcal{H}_i \mid D) = \frac{\mathbb{P}(D \mid \mathcal{H}_i) \cdot \mathbb{P}(\mathcal{H}_i)}{\mathbb{P}(D)} \tag{13}$$

If one supposes that no model is preferred over any other models ($\mathbb{P}(\mathcal{H}_i)$ is constant) then the decision simplifies to finding the model that maximizes

$$\mathbb{P}(D \mid \mathcal{H}_i) = \int_{\mathcal{H}_i} \mathbb{P}(D \mid \mathbf{w}_i, \mathcal{H}_i) \cdot \mathbb{P}(\mathbf{w}_i \mid \mathcal{H}_i) \, d\mathbf{w}_i \tag{14}$$

We make sure that no model parameter is preferred by setting a uniform prior:

$$\mathbb{P}(\mathbf{w}_i \mid \mathcal{H}_i) = 1 / \left(\int_{\mathcal{H}_i} 1 \, d\mathbf{w}_i \right) = 1/\text{Vol}(\mathcal{H}_i) \tag{15}$$

We estimated this integral with Laplace's method by introducing $f(\mathbf{w}_i) := -\frac{1}{|D|} \ln \mathbb{Q}_{\mathbf{w}_i}(D)$, i.e. the cross entropy (measured in nats).

$$\mathbb{P}(D \mid \mathbf{w}_i, \mathcal{H}_i) = \prod_{x \in X} \mathbb{Q}_{\mathbf{w}_i}(x)^{n_x}$$
$$f(\mathbf{w}_i) = - \sum_{x \in X} p_x \cdot \ln \mathbb{Q}_{\mathbf{w}_i}(x) \tag{16}$$

Taking $-\frac{1}{|D|} \ln(\bullet)$ of the evidence amounts to minimizing in i the following quantity:

$$f(\mathbf{w}_i^*) + \frac{1}{|D|} \cdot \ln \text{Vol}(\mathcal{H}_i) + \tag{17}$$
$$\frac{1}{2|D|} \ln \det f''(\mathbf{w}_i^*) + \frac{d}{2|D|} \cdot \ln \frac{|D|}{2\pi}$$

where d is the dimension of $\mathcal{H}_i$ (number of parameters), f'' is the Hessian and $\mathbf{w}_i^* = \arg\min_{\mathbf{w}_i \in \mathcal{H}_i} f(\mathbf{w}_i)$ for a given i. Since the theoretical optimum of $f(\mathbf{w}_i)$ is the entropy of the data ($\ln 2 \cdot \text{H}(D)$), we subtract this quantity from Equation 17 so that the term $f(\mathbf{w}_i)$ becomes the relative entropy (measured in nats) with a theoretical minimum of 0.

We introduce an augmented model to deal with the datapoints where $\mathbb{Q}_{\mathbf{w}_i}(x) = 0$.

$$\overline{\mathbb{Q}}_{\mathbf{w}_i, \mathbf{q}}(x) := \begin{cases} \lambda \mathbb{Q}_{\mathbf{w}_i}(x) & \text{if } \mathbb{Q}_{\mathbf{w}_i}(x) > 0 \\ (1-\lambda)q_x & \text{if } n_x > 0, \mathbb{Q}_{\mathbf{w}_i}(x) = 0 \end{cases} \tag{18}$$

where

$$\lambda = \sum_{x \in X \cap \text{supp}(\mathcal{H}_i)} p_x \qquad \text{covered probability}$$
$$1 - \lambda = \sum_{x \in X \setminus \text{supp}(\mathcal{H}_i)} p_x \qquad \text{uncovered probability}$$

The newly introduced model parameters $\mathbf{q} = (q_x)_{x \in X \setminus \text{supp}(\mathcal{H}_i)}$ are also constrained: they have to be positive and sum up to one, i.e. inside the probability simplex. After finding the optimum of

q and modifying Equation 17 with the auxiliary terms and subtracting the entropy of the data ($\ln 2 \cdot \mathrm{H}(D)$) as discussed above, one gets:

$$-\lambda \cdot \ln \lambda + \sum_{x \in X \cap \mathrm{supp}(\mathcal{H}_i)} p_x \cdot \ln \frac{p_x}{\mathbb{Q}_{\mathbf{w}_i^*}(x)} +$$

$$\frac{1}{|D|} \cdot (\ln \mathrm{Vol}(\mathcal{H}_i) + \ln \mathrm{Vol}(\text{aux. model})) +$$

$$\frac{1}{2|D|} \cdot \ln \det (\text{model Hessian}) +$$

$$\frac{1}{2|D|} \cdot \ln \det (\text{aux. model Hessian}) +$$

$$\frac{d'}{2|D|} \cdot \ln \frac{|D|}{2\pi} \qquad (19)$$

where d' is the original model dimension plus the auxiliary model dimension. One possible use (or abuse) of auxilary parameters would be to directly (nonparametrically) model the low end of the length distribution. But, as we shall see in Section 4, the parametric models actually do better. To see what is going in, let us consider the asymptotic behavior of models.

For sufficiently large corpora ($|D| \to \infty$) all but the first term will be negligible, meaning that the most precise model (in terms of gKL divergence) wins regardless of model size. One way out would be to choose an 'optimum corpus size' (Zipf, 1949), a move that has already drawn strong criticism in Powers (1998) and one that would amount to little more than the addition of an extra hyperparameter to be set heuristically.

Another, more data-driven approach is based on the observation that corpora have *inherent noise*, measurable as the KL divergence between a random subcorpus and its complement (Kornai et al., 2013) both about the same size (half the original). Here we need to take into account the fact that large sentence lengths appear with frequency 1 or 0, so subcorpora D_1 and $D_2 = D \setminus D_1$ will not have the exact same support as the original, and we need to use symmetrized gKL: the **inherent noise** δ_D of a corpus D is $\frac{1}{2}(gKL(D_1, D_2) + gKL(D_2, D_1))$, where D_1 and D_2 are equal size subsets of the original corpus D, and the gKL divergence is measured on their empirical distributions.

δ_D is largely independent of the choice of subsets D_1, D_2 of the original corpus, and can be easily estimated by randomly sampled D_is. To the extent crawl data and classical corpora are sequen-

tially structured (Curran and Osborne, 2002), we sometimes obtain different noise estimates based on random D_i than from comparing the first to the second half of a corpus, the procedure we followed here. In the Minimum Description Length (MDL) setting where this notion was originally developed it is obvious that we need not approximate corpora to a precision better than δ, but in the Bayesian setup that we use here matters are a bit more complicated.

Definition 3.2. *For $\delta > 0$ let*

$$gKL_\delta(\mathbb{P}, \mathbb{Q}) := \max(0, gKL(\mathbb{P}, \mathbb{Q}) - \delta) \quad (20)$$

*For a sample $\mathbb{P}$ with inherent noise δ, a model $\mathbb{Q}$ is called **tolerable** if $gKL_\delta(\mathbb{P}, \mathbb{Q}) = 0$*

If gKL_δ is used instead of gKL in Equation 19 then model size d becomes important. If a model fits within δ then the first term becomes zero and for large $|D|$ values the number of model parameters (including auxiliary parameters) will dominate the evidence. The limiting behavior of our evidence formula, with tolerance for inherent noise, is determined by the following observations:

1. Any tolerable model beats any non-tolerable one.

2. If two models are both tolerable and have different number of model parameters (including auxiliary model), then the one with the fewer parameters wins.

3. If two models are both tolerable and have the same number of parameters, then the model volume and Hessian decides.

An interesting case is when no model can reach the inherent noise – in this case we recover the original situation where the best fit wins, no matter the model size.

4   Results

A single model $\mathcal{H}_i$ fit to some dataset is identified by its *order*, defined as the number of upward steps the random walk can take at once: $1, 2$ or 3, marked by the number before the first decimal; and its *mixture*, a non-empty subset of $\{1, 2, 3, 4, 5\}$ that can appear as k: valency of a single component. For example **1.k1.2.4** marks order 1 and k mixture: $\{1, 2, 4\} \subseteq \{1, 2, 3, 4, 5\}$. Altogether, we trained $3 \times 31 = 93$ locally optimal models for each dataset and compared them with Equation 19, except that gKL_δ is used with the appropriate tolerance.

We computed $\mathbf{w}_i^*$ with a (non-batched) gradient descent algorithm.[4] We used Adagrad with initial learning rate $\eta = 0.9$, starting from uniform p and α values, and iterated until every coordinate of the gradient fell within $\pm 10^{-3}$. The gradient descent typically took $10^2 - 10^3$ iterations to reach a plateau, but about .1% of the models were more sensitive and required a smaller learning rate $\eta = 0.1$ with more (10k) iterations.

4.1 Validation

The model comparison methodology was first tested on artificially generated data. We generated 1M+1M samples of pseudo-random walks with parameters: $p_{-1} = 0.5, p_0 = p_1 = 0.25$ (at most one step upward) and $k = 3$ (no mixture) and obtained the inherent noise and length distribution. The inherent noise was about 3.442e-4 nats. We trained all 93 models and compared them as described above.

The validation data size is $2 \cdot 10^6$ but we also replaced $|D|$ with a hyper-parameter n in Equation 19. This means that we faked the sample to be bigger (or smaller) with the same empirical distribution. We did this with the goal of imitating the 'optimum corpus size' as an adverse effect.

As seen on Table 3 the true model wins. We also tested the case when the true model was simply excluded from the competing models. In this case, the tolerance is needed to ensure a stable result as $n \to \infty$.

1.k3 artificial data	best parameters for various n values					
	1k	10k	100k	1M	10M	1G
with tolerance	3.k1-5	1.k3	1.k3	1.k3	1.k3	1.k3
w/o tolerance	3.k1-5	1.k3	1.k3	1.k3	1.k3	1.k3
w tolerance, -true	3.k1-5	2.k4	2.k4	2.k4	2.k4	2.k4
w/o tolerance, -true	3.k1-5	2.k4	1.k2.3	1.k2.3	1.k2.3	1.k3-5

Table 3: Optimal models for artificially generated data (1.k3) for various n values.

As there are strong conceptual similarities between MDL methods and the Bayesian approach (MacKay, 2003), we also compared the models with MDL, using the same locally optimal parameters as before, but encoding them in bits. To this end we used a technique from (Kornai et al., 2013)

---

[4]You can find all of our code used for training and evaluating at `https://github.com/hlt-bme-hu/SentenceLength`

where all of the continuous model parameters are discretized on a log scale unless the discretization error exceeds the tolerance. The model with the least number of bits required wins if it fits within tolerance. (The constraints are hard-coded in this model, meaning that we re-normalized the parameters after the discretization.) In the artificial test example, the model 1.k3 wins, which is also the winner of the Bayesian comparison. If the true model is excluded, the winner is 1.k2.3. Further MDL results will be discussed in Section 4.4.

4.2 Empirical data

Let us now turn to the natural language corpora summarized in Table 2. Not only are the webcrawl datasets larger than the BNC sections, but they are somewhat noisier and have suspiciously long sentences. To ease the computation, we excluded sentences longer than $1,000$ tokens. This cutoff is always well above the 99.9[th] percentile given in the next to last column of Table 2. The results, summarized in Table 4, show several major tendencies.

First, most of the models (151 out of 174) fit sentence length of the entire subcorpus better than the empirical distribution of the first half would fit the distribution of the second half. When this criterion is *not* met for the best model, i.e. the gKL distance of the model from the data is above the internal noise, the ill-fitting model form is shown in *italics*.

Second, this phenomenon of not achieving tolerable fit is seen primarily (16 out of 29) in the first column of Table 4, corresponding to a radically undersampled condition $n = 1,000$, and (7 out of 29) to a somewhat undersampled condition $n = 10,000$.

Third, and perhaps most important, for sufficiently large n the Bayesian model comparison technique we advocate here actually selects rather simple models, with order 1 (no ditransitives, a matter we return to in Section 5) and only one or two mixture components. We emphasize that 'sufficiently large' is still in the realistic range, one does not have to take the limit $n \to \infty$ to obtain the correct model. The last two columns (gigadata and infinity) always coincide, and in 21 of the 29 corpora the 1M column already yield the same result.

Given that tolerance is generally small, less than 0.66 bits even in our noisiest corpus (BNC-K), we didn't expect much change if we perform the model comparison without using Equation 20. Unsurprisingly, if we reward every tiny improvement in divergence, we get more models (159 out of 174) within the tolerable range – those outside the tolerance limit are again given in italics in Table 6. But we pay a heavy price in model complexity: the best models (in the last two columns) are now often second order, and we have to countenance a hyperparameter n which matters (e.g. for Polish).

dataset	best parameters for various n values					
	1k	10k	100k	1M	1G	∞
BNC-A	*3.k1-5*	*3.k2-5*	1.k4.5	1.k4.5	1.k4.5	1.k4.5
BNC-B	*3.k1-5*	*3.k1-5*	1.k1.5	1.k1.5	1.k1.5	1.k1.5
BNC-C	3.k2-5	3.k2-5	3.k2-5	1.k1.4	1.k1.4	1.k1.4
BNC-D	3.k2.3.5	3.k2.3.5	3.k2.3.5	1.k2	1.k2	1.k2
BNC-E	*3.k1.3-5*	*3.k1.3-5*	1.k2.5	1.k2.5	1.k2.5	1.k2.5
BNC-F	3.k3.4.5	3.k3.4.5	3.k3.4.5	1.k3	1.k3	1.k3
BNC-G	3.k1-5	3.k1-5	1.k2.5	1.k2.5	1.k2.5	1.k2.5
BNC-H	*3.k2.4.5*	3.k3.4.5	1.k4	1.k4	1.k4	1.k4
BNC-J	3.k2.3.4	3.k2.3.4	3.k2.5	1.k2	1.k2	1.k2
BNC-K	3.k1-5	3.k1-5	1.k2	1.k2	1.k2	1.k2
UMBC	*3.k1.3-5*	*3.k1.3-5*	1.k2.5	1.k2.5	1.k2.5	1.k2.5
Catalan	*3.k2-5*	*3.k2-5*	1.k2.5	1.k2.5	1.k2.5	1.k2.5
Croatian	3.k3.4.5	3.k3.4.5	1.k2.5	1.k2.5	1.k2.5	1.k2.5
Czech	*3.k4.5*	3.k1.3.5	1.k2.5	1.k2.5	1.k2.5	1.k2.5
Danish	*3.k1-5*	3.k1.3.5	1.k2.5	1.k2.5	1.k2.5	1.k2.5
Dutch	*3.k1-5*	*3.k3.4.5*	1.k2.5	1.k2.5	1.k2.5	1.k2.5
Finnish	*3.k1.3.5*	1.k2.4	1.k2.4	1.k2.4	1.k2.4	1.k2.4
Indonesian	3.k1-5	3.k1-5	1.k2.5	1.k2.5	1.k2.5	1.k2.5
Lithuanian	*3.k2.3.4*	*3.k2.3.4*	1.k2.3	1.k2.3	1.k2.3	1.k2.3
Bokmål	3.k2.4.5	3.k2.4.5	1.k2.5	1.k2.5	1.k2.5	1.k2.5
Nynorsk	*3.k1-5*	1.k2.5	1.k2.5	1.k2.5	1.k2.5	1.k2.5
Polish	3.k2-5	3.k2-5	3.k2-5	3.k2-5	1.k2.5	1.k2.5
Portuguese	3.k2.3.5	3.k2.3.5	1.k2	1.k2	1.k2	1.k2
Romanian	3.k1.3-5	3.k1.3-5	1.k5	1.k5	1.k5	1.k5
Serbian.sh	*3.k1.2.4.5*	3.k2.3.5	1.k2.5	1.k2.5	1.k2.5	1.k2.5
Serbian.sr	*3.k2-5*	3.k2.3.4	1.k2.5	1.k2.5	1.k2.5	1.k2.5
Slovak	*3.k2.4.5*	3.k2-5	1.k2.5	1.k2.5	1.k2.5	1.k2.5
Spanish	*3.k2.4.5*	1.k2.3	1.k2.3	1.k2.3	1.k2.3	1.k2.3
Swedish	1.k2.4	1.k2.4	1.k2.4	1.k2.4	1.k2.4	1.k2.4

Table 4: Optimal models with tolerance for inner noise. *Ill-fitting models* are marked with italics.

4.3 Previous models

We also compared previous or baseline sentence length models with our new model. The hyper-parameters of the *bins* model are the bins themselves. The distribution over the bins are the continuous model-parameters. For m bins: $[1, b_1), [b_1, b_2), \ldots [b_{m-1}, \infty)$, the probability distribution $\mathbb{P}(b_i \leq X < b_{i+1}) = q_i$ is to be optimized. This model has $m - 1$ free parameters (model dimension) and its model volume is the volume of a probabilistic m-simplex. No auxiliary model is required.

We also trained[5] and compared Sichel's model (Equation 1) with our method. In this case α and θ are the model-parameters and γ was a non-trained hyper-parameter. In Sichel (1974) it was fixed $\gamma = -\frac{1}{2}$, we trained $\gamma \in \{-0.5, -0.4\}$, the higher γ value was usually better. Again no auxiliary model was needed.

dataset	Sichel	binned	randwalk	δ
BNC-A	*3.554e-2*	*1.489e-2*	4.409e-4	9.847e-4
BNC-B	*6.212e-2*	*1.274e-2*	7.215e-3	7.741e-3
BNC-C	*4.861e-2*	*1.431e-2*	6.989e-3	9.494e-3
BNC-D	*9.917e-2*	8.387e-2	5.945e-2	8.510e-2
BNC-E	*6.976e-2*	*2.251e-2*	4.353e-3	5.000e-3
BNC-F	*3.153e-2*	2.196e-2	2.270e-2	2.630e-2
BNC-G	*2.598e-2*	*1.495e-2*	5.762e-3	9.199e-3
BNC-H	*4.765e-2*	3.265e-2	3.106e-2	3.385e-2
BNC-J	3.048e-2	6.854e-2	2.946e-2	7.940e-2
BNC-K	6.583e-2	1.388e-1	3.899e-2	2.134e-1
UMBC	6.584e-2	2.615e-2	1.390e-3	2.442e-3
Catalan	*1.389e-1*	6.102e-2	9.382e-4	1.751e-3
Croatian	*1.131e-1*	4.604e-2	2.063e-3	5.616e-3
Czech	5.857e-2	3.687e-2	2.563e-3	5.147e-3
Danish	*1.618e-1*	3.072e-2	2.772e-3	7.557e-3
Dutch	4.232e-1	3.447e-2	1.391e-3	2.408e-3
Finnish	9.968e-2	2.830e-2	1.659e-3	1.946e-3
Indonesian	*2.159e-1*	5.017e-2	1.390e-3	1.231e-2
Lithuanian	-	3.113e-2	6.637e-4	1.184e-3
Bokmål	-	3.332e-2	3.515e-3	3.564e-3
Nynorsk	-	2.830e-2	3.757e-3	3.946e-3
Polish	-	4.078e-2	1.518e-3	8.508e-3
Portuguese	-	5.133e-2	4.514e-2	4.973e-2
Romanian	-	6.539e-2	1.579e-2	2.338e-2
Serbian.sh	-	4.676e-2	1.346e-3	4.531e-3
Serbian.sr	-	*1.389e-1*	6.971e-3	7.189e-3
Slovak	-	4.344e-2	2.184e-3	2.572e-3
Spanish	-	6.501e-2	7.718e-4	8.365e-4
Swedish	-	2.652e-2	2.310e-3	2.526e-3

Table 5: Best of the models and their fit. *Ill-fitting models* are marked with italics.

As can be seen, the fit is always improved (on the average by 40%) from the mixture Poisson to the binned model, and the random walk model further improves from the binned (on the average by 70%). More important, the mixture Poisson model never, the binned model rarely, but the random walk model always approximates the data better than its inner noise. Altogether the random walk models always outperforms the other two, but not always for the same reason. In the case of bins, the fit was poor and only the fine-grained bins per-

---

[5]Optimizing the mixture Poisson coefficients took orders of magnitude more time than optimizing the other models. The difficulties come from computing the derivatives of Bessel functions. At the time of going to press still about a third of the values are missing – by the time of the meeting these will be published at `https://github.com/hlt-bme-hu/SentenceLength`

formed within inherent noise. Note that none of our parametric models use mode than 11 parameters, which makes only systems with 12 or fewer bins competitive.

In case of Equation 1, Sichel already mentions that the fit is satisfactory only with binned probabilities, i.e. on a dumbed down distribution with 4-5 data points aggregated into one. This classic model has only 2 parameters, which would make it very competitive for large inherent noise or small data size, but neither is the case here.

4.4 MDL approach

Finally, let us consider the MDL results given in Table 7. These are often (9 out of 29 subcorpora) consistent with the results obtained using Equation 20, but never with those obtained without considering inherent noise to be a factor. Remarkably, we never needed more than 6 bits quantization, consistent with the general principles of Google's TPUs (Jouppi et al., 2017) and is in fact suggestive of an even sparser quantization regime than the eight bits employed there.

For a baseline, we discretized the naive (non-parametric) model in the same way. Not only does the quantization require on the average two bits more, but we also have to countenance a considerably larger number of parameters to specify the distribution within inherent noise, so that the random walk model offers a size savings of at least 95.3% (BNC-A) to 99.7% (Polish).

With the random walk model, the total number of bits required for characterizing the most complex distributions (66 for BNC-A and 60 for Spanish) appears to be more related to the high consistency (low internal noise) of these corpora than to the complexity of the length distributions.

5 Conclusion

At the outset of the paper we criticized the standard mixture Poisson length model of Equation 1 for lack of a clear genesis – there is no obvious candidate for 'arrivals' or for the mixture. In contrast, our random walk model is based on the suggestive idea of total valency 'number of things you want to say', and we see some rather clear methods for probing this further.

First, we have extensive lexical data on the valency of individual words, and know in advance that e.g. color adjectives will be dependent on nouns, while relational nouns such as *sister* can

dataset	best parameters for various n values				
	1k	10k	100k	1M	1G
BNC-A	*3.k1-5*	1.k4.5	1.k4.5	1.k1-5	1.k1-5
BNC-B	*3.k1-5*	1.k2.3.5	2.k4.5	2.k4.5	2.k4.5
BNC-C	3.k2-5	1.k2.4.5	1.k2.4.5	1.k2.4.5	1.k2.4.5
BNC-D	3.k3.4	1.k2.5	2.k2.5	2.k2.5	2.k2.5
BNC-E	*3.k1.3-5*	1.k4.5	1.k4.5	1.k4.5	1.k4.5
BNC-F	3.k3-5	1.k2.4.5	1.k2.4.5	1.k2.4.5	1.k2.4.5
BNC-G	3.k1-5	1.k4.5	1.k2.4.5	1.k2.4.5	2.k2.4.5
BNC-H	3.k3-5	1.k4.5	2.k2.4.5	2.k2.4.5	2.k2.4.5
BNC-J	3.k1-5	1.k2.4.5	1.k2.4.5	1.k2.4.5	1.k2.4.5
BNC-K	3.k2-5	3.k2-5	1.k2.4.5	1.k2.4.5	1.k2.4.5
UMBC	*3.k1.3-5*	1.k2.4	1.k2.4.5	1.k2.4.5	1.k2.4.5
Catalan	*3.k2-5*	*3.k2-5*	1.k2.4	1.k1.3-5	1.k1.3-5
Croatian	3.k3-5	1.k2.3	1.k2.3	1.k3-5	1.k3-5
Czech	3.k2-5	3.k3-5	1.k2.3	1.k1.3-5	1.k1.3-5
Danish	*3.k1-5*	1.k2.3	1.k1.2.4.5	1.k1.2.4.5	3.k2-5
Dutch	*3.k1-5*	1.k2.4	1.k3.4	1.k1-5	1.k1-5
Finnish	*3.k1.3.5*	1.k1.3.4	1.k1.3.4	1.k1.3-5	1.k1.3-5
Indonesian	3.k1-5	1.k3.5	1.k3-5	1.k3-5	1.k3-5
Lithuanian	*3.k2.3.4*	1.k2.3	1.k2-5	1.k2-5	1.k2-5
Bokmål	3.k2.4.5	3.k2.4.5	1.k1.3-5	1.k1.3-5	1.k1.3-5
Nynorsk	*3.k1-5*	1.k2.4.5	1.k1-5	1.k1-5	1.k1-5
Polish	3.k2-5	3.k2-5	1.k1.4.5	1.k2-5	1.k2-5
Portuguese	3.k2.4.5	1.k2.3	1.k3.4	1.k3.4	1.k3.4
Romanian	3.k2.4.5	1.k2.4	1.k2.3.4	1.k2.3.4	1.k2.3.4
Serbian.sh	*3.k1.2.4.5*	1.k2.4	1.k3.4	1.k2-5	1.k2-5
Serbian.sr	*3.k2-5*	1.k4.5	1.k4.5	1.k4.5	1.k4.5
Slovak	*3.k2.4.5*	1.k2.3	1.k1.3-5	1.k1.3-5	1.k1.3-5
Spanish	*3.k2.4.5*	1.k2.3	1.k1.3.5	1.k1.3.5	1.k1.3.5
Swedish	1.k2.3	1.k2.3	1.k1-5	1.k1-5	1.k1-5

Table 6: Optimal models without tolerance. *Ill-fitting models* are marked with italics.

bring further nouns or NPs. Combining the lexical knowledge with word frequency statistics is somewhat complicated by the fact that a single word form may have different senses with different valency frames, but these cause no problems for a statistical model that convolves the two distributions.

Second, thanks to Universal Dependencies[6] we now have access to high quality dependency treebanks where the number of dependencies running between words $w_1, \ldots, w_k$ and $w_{k+1} \ldots w_n$, the y coordinate of our random walk at k, can be explicitly tracked. Using these treebanks, we could perform a far more detailed analysis of phrase or clause formation than we attempted here, e.g. by systematic comparison of the learned p_1 and p_2 values with the observable proportion of intransitive and transitive verbs and relational nouns. Ditransitives are rare (in fact they usually make up less than 2% of the verbs) and we think these can be eliminated entirely (Kornai, 2012) without loss

---

[6]http://universaldependencies.org

dataset	mq	nq	tb	opt	% size
BNC-A	6	7	66	1.k4.5	4.69
BNC-B	4	5	40	2.k2.5	4.65
BNC-C	3	5	36	1.k1.2.4	3.32
BNC-D	2	3	6	1.k2	1.29
BNC-E	4	5	32	1.k2.5	3.56
BNC-F	2	5	16	1.k2.5	1.13
BNC-G	3	5	24	1.k1.2	2.45
BNC-H	2	4	16	1.k2.5	1.36
BNC-J	2	4	6	1.k2	0.51
BNC-K	2	3	6	1.k2	0.63
UMBC	4	7	44	1.k4.5	0.88
Catalan	5	7	40	1.k2.5	0.57
Croatian	4	6	32	1.k2.4	0.53
Czech	3	6	24	1.k2.5	0.41
Danish	3	6	24	1.k2.4	0.41
Dutch	5	7	40	1.k2.5	0.57
Finnish	4	7	48	1.k1.2.3	0.69
Indonesian	4	5	32	1.k2.4	0.66
Lithuanian	4	7	32	1.k2.3	0.46
Bokmål	3	7	30	2.k2.5	0.43
Nynorsk	4	6	32	1.k2.3	1.14
Polish	2	5	16	1.k2.5	0.32
Portuguese	3	5	18	1.k4	0.36
Romanian	3	5	24	1.k1.2	0.48
Serbian.sh	4	6	32	1.k2.4	0.53
Serbian.sr	4	5	32	1.k2.5	0.64
Slovak	5	6	40	1.k2.3	0.67
Spanish	6	7	60	1.k3.4	0.86
Swedish	5	7	40	1.k2.3	0.57

Table 7: Optimal models with MDL comparison (with tolerance). mq: Model quantization bits. nq: naive/nonparametric quantization bits. tb: total bits. opt: optimal model configuration. %size: size of random walk model as percentage of size of nonparametric model.

of generality. The same kind of analysis could be attempted for other grammatical formalisms like type-logical grammars, which make tracking the open arguments an even more attractive proposition, but unfortunately these lack large parsed corpora. Another significant issue with formalisms other than UD is that the cross-linguistic breadth of parsed corpora is minute – do we want to base general conclusions of the type attempted here, linking predicate/argument structure to sentence length, on English alone?

Third, we can extend the analysis in a typologically sound manner to morphologically more complex languages. Using a morphologically analyzed Hungarian corpus (Oravecz et al., 2014) we measured the per-word morpheme distribution and per-sentence word distribution. We found that the random sum of 'number of words in a sentence' independent copies of 'number of morphemes in a word' estimates the per-sentence morpheme dis-

tribution within inherent noise. To the extent these results can be replicated for other morphologically complex languages (again UD morphologies[7] offer the best testbed, though a lot remains to be done for ensuring homogeneity) problems like six-word 'I can give you a ride' versus one-word *elvihetlek* disappear.

Another avenue of research alluded to above would be the study of subject- and object-control verbs and infinitival constructions, where single nouns or NPs can fill more than one open dependency. This would complicate the calculations in Equation 5 in a non-trivial way. We plan to extend our mathematical model in a future work, but it should be clear from the foregoing that sentences exhibiting these phenomena are so rare as to render unlikely any prospect of improving the statistical model by means of accounting for these. This is not to say that control phenomena are irrelevant to grammar – but they are likely 'within the noise' for statistical length modeling.

One of the authors (Kornai and Tuza, 1992) already suggested that the number of dependencies open at any given point in the sentence must be subject to limitations of short-term memory (Miller, 1956) – this may act as a reflective barrier that keeps asymptotic sentence length smaller than the pure random walk model would suggest. In particular, Bernoulli and other well-known models predict exponential decay at the high end, whereas our data shows polynomial decay proportional to n^{-C}, with C somewhere around 4 (in the $3-5$ range). This is one area where our corpora are too small to draw reliable conclusions, but overall we should emphasize that corpora already collected (and in the case of UD treebanks, already analyzed) offer a rich empirical field for studying sentence length phenomena, and the model presented here makes it possible to use statistics to shed light on the underlying grammatico-semantic structure.

Acknowledgments

The presentation greatly benefited from the remarks of the anonymous reviewers. Research partially supported by National Research, Development and Innovation Office (NKFIH) grants #120145: Deep Learning of Morphological Structure and NKFIH grant #115288: Algebra and al-

---

[7] https://universaldependencies.org/u/ overview/morphology.html

gorithms as well as by National Excellence Programme 2018-1.2.1-NKP-00008: Exploring the Mathematical Foundations of Artificial Intelligence. A hardware grant from NVIDIA Corporation is gratefully acknowledged. GNU parallel was used to run experiments (Tange, 2011).

References

Yossi Adi, Einat Kermany, Yonatan Belinkov, Ofer Lavi, and Yoav Goldberg. 2017. Fine-grained analysis of sentence embeddings using auxiliary prediction tasks. In *Proceedings of International Conference on Learning Representations*.

Alexis Conneau, Germán Kruszewski, Guillaume Lample, Loïc Barrault, and Marco Baroni. 2018. What you can cram into a single \\\$&!#* vector: Probing sentence embeddings for linguistic properties. In *Proceedings of the 56th Annual Meeting of the Association for Computational Linguistics (Volume 1: Long Papers)*, pages 2126–2136. Association for Computational Linguistics.

James R. Curran and Miles Osborne. 2002. A very very large corpus doesn't always yield reliable estimates.

Norman P. Jouppi, Cliff Young, Nishant Patil, David Patterson, and et. al. 2017. In-datacenter performance analysis of a tensor processing unit. In *Proceedings of ISCA '17*.

András Kornai. 2012. Eliminating ditransitives. In Ph. de Groote and M-J Nederhof, editors, *Revised and Selected Papers from the 15th and 16th Formal Grammar Conferences*, LNCS 7395, pages 243–261. Springer.

András Kornai and Zsolt Tuza. 1992. Narrowness, pathwidth, and their application in natural language processing. *Discrete Applied Mathematics*, 36:87–92.

András Kornai, Attila Zséder, and Gábor Recski. 2013. Structure learning in weighted languages. In *Proceedings of the 13th Meeting on the Mathematics of Language (MoL 13)*, pages 72–82, Sofia, Bulgaria. Association for Computational Linguistics.

David J.C. MacKay. 2003. *Information Theory, Inference, and Learning Algorithms*. Cambridge University Press.

T.C. Mendenhall. 1887. The characteristic curves of composition. *Science*, 11:237–249.

Jason Merchant. 2001. *The Syntax of Silence: Sluicing, Islands, and the Theory of Ellipsis*. Oxford University Press.

George A. Miller. 1956. The magical number seven, plus or minus two: some limits on our capacity for processing information. *Psychological Review*, 63:81–97.

George A. Miller. 1957. Some effects of intermittent silence. *American Journal of Psychology*, 70:311–314.

Csaba Oravecz, Tamás Váradi, and Bálint Sass. 2014. The Hungarian Gigaword Corpus. In *Proceedings of the Ninth International Conference on Language Resources and Evaluation (LREC-2014)*. European Language Resources Association (ELRA).

David M.W. Powers. 1998. Applications and explanations of Zipf's law. In D.M. W. Powers, editor, *NEMLAP3/CONLL98: New methods in language processing and Computational natural language learning*, pages 151–160. ACL.

H.S. Sichel. 1974. On a distribution representing sentence length in written prose. *Journal of the Royal Statistical Society Series A*, 137(1):25–34.

Andreas Stolcke, Jing Zheng, Wen Wang, and Victor Abrash. 2011. Srilm at sixteen: Update and outlook. In *Proceedings of IEEE Automatic Speech Recognition and Understanding Workshop*, volume 5.

O. Tange. 2011. Gnu parallel - the command-line power tool. *;login: The USENIX Magazine*, 36(1):42–47.

W.C. Wake. 1957. Sentence-length distributions of Greek authors. *Journal of the Royal Statistical Society Series A*, 120:331–346.

C.B. Williams. 1944. A note on the statistical analysis of sentence-length as a criterion of literary style. *Biometrika*, 31:356–361.

G. Udny Yule. 1939. On sentence-length as a statistical characteristic of style in prose: with applications to two cases of disputed authorship. *Biometrika*, 30:363–390.

G. Udny Yule. 1944. *The Statistical Study of Literary Vocabulary*. Cambridge University Press.

George K. Zipf. 1949. *Human Behavior and the Principle of Least Effort*. Addison-Wesley.

A Appendix

Theorem A.1. *Let us define f as $x = \frac{f(x)}{F(f(x))}$ with $F(0) > 0$, then*

$$\left[x^i\right](f(x))^k = \frac{k}{i}[x^{i-k}]F^i(x) \qquad (21)$$

Proof. By Lagrange–Bürmann formula with composition function $H(x) = x^k$. □

Theorem A.2. *In the Bayesian evidence if both the model and parameter a priori is uniform, then*

$$\mathbb{P}(\mathcal{H}_i \mid D) = \frac{\mathbb{P}(D \mid \mathcal{H}_i) \cdot \mathbb{P}(\mathcal{H}_i)}{\mathbb{P}(D)} \propto f(\mathbf{w}_i^*) +$$

$$\frac{1}{n} \cdot \ln \mathrm{Vol}(\mathcal{H}_i) + \frac{1}{2n} \ln \det f''(\mathbf{w}_i^*) + \frac{d}{2n} \cdot \ln \frac{n}{2\pi}$$

where $f(\mathbf{w}_i)$ is the cross entropy of the measured and the modeled distributions. See Equation 17.

If the augmented model (18) is used, then Equation 19 follows.

Proof.

$$\mathbb{P}(D \mid \mathcal{H}_i) \overset{\text{uniform a priori}}{=}$$

$$\int \mathbb{P}(D \mid \mathbf{w}_i, \mathcal{H}_i) \cdot \frac{1}{\mathrm{Vol}(\mathcal{H}_i)} \, \mathrm{d}\mathbf{w}_i =$$

$$\frac{1}{\mathrm{Vol}(\mathcal{H}_i)} \cdot \int \prod_{x \in X} \mathbb{Q}_{\mathbf{w}_i}(x)^{n_x} \, \mathrm{d}\mathbf{w}_i =$$

$$\frac{\int \exp\left\{ -n \cdot \overbrace{\left(-\sum_{x \in X} \frac{n_x}{n} \cdot \ln \mathbb{Q}_{\mathbf{w}_i}(x) \right)}^{f(\mathbf{w}_i)} \right\} \mathrm{d}\mathbf{w}_i}{\mathrm{Vol}(\mathcal{H}_i)}$$

Using Laplace method:

$$\approx \frac{1}{\mathrm{Vol}(\mathcal{H}_i)} \cdot e^{-n \cdot f(\mathbf{w}_i^*)} \cdot \frac{\left(\frac{2\pi}{n}\right)^{\frac{d}{2}}}{\sqrt{\det f''(\mathbf{w}_i^*)}}$$

Taking $-\frac{1}{n}\ln(\bullet)$ for scaling (does not effect the relative order of the models):

$$\frac{1}{n}\ln \mathrm{Vol}(\mathcal{H}_i) + f(\mathbf{w}_i^*) + \frac{1}{2n}\ln \det f''(\mathbf{w}_i^*) +$$

$$\frac{d}{2n} \cdot \ln\left(\frac{n}{2\pi}\right)$$

As for the augmented model, the model parameters are the concatenation of the original parameters and the auxiliary parameters. Thus the overall Hessian is the block-diagonal matrix of the original and the auxiliary Hessian. Similarly, the overall model volume is the product of the original and the auxiliary volume. Trivially, the logarithm of product is the sum of the logarithms.

Since the auxiliary model can fit the uncovered part perfectly: $p_x = (1 - \lambda) \cdot q_x$ on $x \notin \mathrm{supp}\,\mathcal{H}_i$. See (18) for that λ is the covered probability of the sample.

$$\mathbb{P}(D \mid \mathcal{H}_i') = - \sum_{x \in X \setminus \mathrm{supp}(\mathcal{H}_i)} p_x \cdot \ln p_x$$

$$- \sum_{x \in X \cap \mathrm{supp}(\mathcal{H}_i)} p_x \cdot \ln\left(\lambda \cdot \mathbb{Q}_{\mathbf{w}_i^*}(x) \right) +$$

$$\frac{1}{n} \cdot \left(\ln \mathrm{Vol}(\mathcal{H}_i) + \ln \mathrm{Vol}(\text{aux. model}) \right) +$$

$$\frac{1}{2n} \cdot \ln \det (\text{model Hessian}) +$$

$$\frac{1}{2n} \cdot \ln \det (\text{aux. model Hessian}) +$$

$$\frac{d'}{2n} \cdot \ln \frac{n}{2\pi} \qquad (22)$$

where d' is the overall parameter number.

Further, if one subtracts the entropy of the sample then only the first two term is changed compared to Equation 22 and Equation 19 follows.

$$\sum_{x \in X} p_x \cdot \ln p_x - \sum_{x \in X \setminus \mathrm{supp}(\mathcal{H}_i)} p_x \cdot \ln p_x$$

$$- \sum_{x \in X \cap \mathrm{supp}(\mathcal{H}_i)} p_x \cdot \ln\left(\lambda \cdot \mathbb{Q}_{\mathbf{w}_i^*}(x) \right) =$$

$$\sum_{x \in X \cap \mathrm{supp}(\mathcal{H}_i)} p_x \cdot \ln \frac{p_x}{\lambda \cdot \mathbb{Q}_{\mathbf{w}_i^*}(x)} =$$

$$\sum_{x \in X \cap \mathrm{supp}(\mathcal{H}_i)} p_x \cdot \left(\ln \frac{p_x}{\mathbb{Q}_{\mathbf{w}_i^*}(x)} + \ln \frac{1}{\lambda} \right) =$$

$$\lambda \cdot (-\ln \lambda) + \sum_{x \in X \cap \mathrm{supp}(\mathcal{H}_i)} p_x \cdot \ln \frac{p_x}{\mathbb{Q}_{\mathbf{w}_i^*}(x)}$$

q.v. Definition 3.1. $\qquad \square$